Polymers

Properties and Applications 8

W. Lincoln Hawkins

Polymer Degradation and Stabilization

With 33 Figures

Springer-Verlag
Berlin Heidelberg New York Tokyo 1984

Dr. W. Lincoln Hawkins
Materials Consultant
26 High Street
Montclair, NJ 07042/USA

Editor:
Professor Dr. H. James Harwood
Institute of Polymer Science
University of Akron
Akron, OH 44325/USA

This volume continues the series *Chemie, Physik und Technologie der Kunststoffe in Einzeldarstellungen,* which is now entitled *Polymers/Properties and Applications.*

ISBN-13: 978-3-642-69378-6   e-ISBN-13: 978-3-642-69376-2
DOI: 10.1007/978-3-642-69376-2

Library of Congress Cataloging in Publication Data:
Hawkins, W. Lincoln, 1931 – Polymers degradation and stabilization. (Polymers; 8) Includes index. 1. Polymers and polymerization–Deterioration. 2. Stabilizing agents. I. Title. II. Series.
QD380.H38 1984   620.1′920422   83-16699

Typesetting, Printing, and Bookbinding: Brühlsche Universitätsdruckerei, D-6300 Giessen
2152/3020-543210

To my wife, Lilyan
and our sons, Gordon and Philip

Preface

The development of polymers as an important class of material was inhibited at the first by the premature failure of these versatile compounds in many applications. The deterioration of important properties of both natural and synthetic polymers is the result of irreversible changes in composition and structure of polymers molecules. As a result of these reactions, mechanical, electrical and/or aesthetic properties are degraded beyond acceptable limits. It is now generally recognized that stabilization against degradation is necessary if the useful life of polymers is to be extended sufficiently to meet design requirements for long-term applications.

Polymers degrade by a wide variety of mechanisms, several of which affect all polymers through to varying degree. This monograph will concentrate on those degradation mechanisms which result from reactions of polymers with oxygen in its various forms and which are accelerated by heat and/or radiation. Those stabilization mechanisms are discussed which are based on an understanding of degradation reaction mechanisms that are reasonably well established.

The stabilization of polymers is still undergoing a transition from an art to a science as mechanisms of degradation become more fully developed. A scientific approach to stabilization can only be approached when there is an understanding of the reactions that lead to degradation. Stabilization against biodegradation and burning will not be discussed since there is not a clear understanding of how polymers degrade under these conditions.

It is the purpose of this monograph to provide a critical review of mechanisms for polymer degradation and stabilization that will be of practical value to students and industrial chemists.

Montclair, October 1983   W. Lincoln Hawkins

Table of Contents

A) Introduction

The degradation of polymers under normal use conditions is a major factor limiting application of these remarkable and versatile materials. Without exception, all polymers are degraded eventually within the environment to which they are exposed during their life cycle. Though there is considerable variation between polymers in their resistance to degradation, eventually important properties of every polymer are affected adversely as those chemical reactions responsible for degradation proceed. Loss in mechanical strength, dielectric quality, and aesthetic appearance often lead to failure of polymeric materials – before reaching the required service life.

Stabilization is required to extend the useful life of most polymers. Compounding with selected additives (stabilizers) is the favored method for improving stability. Alteration of polymer molecules to provide more resistant structures has a limited role in stabilization. In altering the structure of a polymer to obtain a required level of stability, every other important property of that polymer must be held within specified limits. This is difficult at best and usually impossible to accomplish. This monograph will concentrate on the use of additives to improve stability. Only limited examples of structure modification will be discussed, in particular the elimination of imperfections in molecular structure and incorporation of stabilizer moieties into polymer molecules.

Development of the art of stabilization into a science took place rapidly after Hoffman reported in 1861 [1] that degradation of natural rubber was accompanied by absorption of oxygen. Inhibition of the degradation process by addition of phenols or amines was observed by many of the earlier investigators [2–4]. Over the course of several decades, it was finally established that the degradation of natural rubber occurred by a free-radical-initiated [5], chain mechanism [6–8]. The misleading hypothesis that stabilizers functioned as oxygen scavengers was abandoned as it was recognized that protection was provided by those additives which in some way inhibited one or more of the reactions occurring in the polymer. The term "anti-ger" or "anti-oxygen" was then abandoned in favor of the modern term "antioxidant" to describe the class of stabilizers which protects against oxygen-induced degradation. As the science of stabilization developed, several other types of stabilizers appeared, each functioning at a specific stage in the degradation mechanism. Stabilizers were developed to protect against ultraviolet-induced degradation, ozone attack, and ionizing radiation.

The key point which evolved from the early history of polymer stabilization was simply that a scientific approach could only be developed after the mechanism responsible for degradation had been established. Because of the complexity of polymer reactions, much of the early mechanistic studies were developed through the use of model compounds. Despite the obvious fact that a perfect model for a complex polymer molecule is difficult to devise, experiments with simple compounds made an invaluable contribution to elucidation of degradation mechanisms for several polymers. Studies on the oxidation of model hydrocarbons at the Natural Rubber Producers' Research Association [9, 10] provided the basic understanding for the degradation of natural rubber as well as the polyolefins. More recently, model compound studies have contributed to understanding of the degradation mechanism for other polymers, e.g. poly(vinyl chloride).

This review will deal first with those degradation mechanisms for which there is significant experimental support. Stabilization will then be discussed as the science has developed during the past several decades both in thermal and in photodegradation. A detailed discussion of stabilization against burning is considered to be beyond the scope of this review, but a short summary is included.

References

1. Hoffman, A.W.: J. Chem. Soc. 87 (1861)
2. Ostwald, W., Ostwald, W.: German Patent 221,320 (1908)
3. Moureu, C., Dufraisse, C.: Bull. Soc. Chim. Fr. *31*, 1152 (1922)
4. Hoffman, A.W.: History and Use of Materials which Improve Aging, in: The Chemistry and Technology of Rubber (ed) Davis, C.C. pp. 414, New York: Reinhold 1937
5. Bäckström, H.L.J.: Z. Phys. Chem. B., *25*, 122 (1954)
6. Bodenstein, M.: Z. Phys. Chem., *84*, 329 (1913)
7. Christiansen, J.A.: J. Phys. Chem., *28*, 145 (1924)
8. Bäckström, M.: J. Amer. Chem. Soc., *49*, 1460 (1927)
9. Bolland, J.L.: Quart. Rev. (London), *3*, 1 (1949)
10. Bateman, L.: Quart. Rev. (London), *8*, 147 (1954)

B) Polymer Degradation

Table B-1. Bond dissociation energies

Bond	Dissociation energy Kcal/mol
$-\overset{\mid}{C}---\overset{\mid}{C}-$	83
$-\overset{\mid}{C}---H$	91
$-\overset{\mid}{C}H---H$	95
$-CH_2---H$	98
$-CH_2---O-$	106
$-CH_2---F$	93
$-\overset{\mid}{C}---N-$	82
$-\overset{\mid}{Si}---O-$	106
$-\overset{\mid}{C}H---Cl$	78

ogy has been used by many investigators in the study of thermal degradation or pyrolysis. This research has been important in determining the relative strength of bonds in polymer molecules. As energy is absorbed and distributed through the molecules, a point is reached at which the energy concentrated at one bond in the molecule exceeds its dissociation energy. When this point is reached, the bond ruptures and unless it can reform, an irreversible chemical change will take place. It is important, therefore, to know the dissociation energy of those bonds which occur in polymer molecules. Typical bond dissociation energies are shown in Table B-1.

By rigid definition, pyrolysis of a polymer is thermal degradation in the complete absence of any external reactant. Thermal degradation, however, usually results from the combined effects of pyrolysis and thermal oxidation. Hydrolytic reactions due to traces of entrapped moisture may also contribute, particularly in condensation polymers. Isolation of pyrolytic reactions by the total exclusion of oxygen or water can only occur in a total vacuum or in a completely inert and dry atmosphere. These conditions have been closely approximated by careful experimenters, and their results can be interpreted as due only to pyrolysis reactions. In practice, however, thermal degradation usually takes place in a limited oxygen atmosphere. Traces of moisture may also be present.

Most processing equipment restricts access of oxygen to the molten polymer and pyrolysis may be the predominating degradation reaction during processing. However, the contribution of thermal oxidation cannot be ignored. In the following discussion of thermal degradation, those reactions which take place in a highly restricted oxygen atmosphere will be considered as essentially pyrolytic degradation.

The earliest experiment on record of thermal degradation in an atmosphere of limited oxygen content is credited to Williams[3] who in 1860 demonstrated that isoprene is formed when natural rubber is heated. This led eventually to Staudinger's "giant molecule" concept for the structure of polymers. There then followed a long sequence of research papers[4–6] dealing with the pyrolysis of natural rubber and synthetic polymers. As a result of the increasing number of applications for polymers and polymer blends under conditions of exposure to heat in a restricted oxygen atmosphere, research in this important field is continuing. Before discussing current research, however, some of the fundamental reactions of pyrolysis should be reviewed.

Polymers pyrolyze by one of three general mechanisms or by combinations of two or more of these mechanisms. Random scission, which is the predominant reaction in polyolefins, occurs through scission of bonds along the backbone chain. As the term implies, scission is a random event, and polymer molecules are first broken into large macroradicals. There is a rapid decrease in molecular weight and almost no monomer is formed in the early stages.

Depolymerization is the second general mechanism of pyrolysis. This reaction is usually initiated at chain ends. Monomer units are split off sequentially. Since only a small fraction of the molecules react at the beginning, there is very little change in molecular weight. Once depolymerization has been initiated in a molecule, the reaction proceeds until that molecule has depolymerized completely. Polyacetals and poly(methyl methacrylate) pyrolyze by depolymerization.

Poly(vinyl chloride) (PVC) pyrolyzes by the third general mechanism. This mechanism is characterized by the elimination of a low-molecular-weight fragment – other than monomer. The splitting off of hydrogen chloride from PVC is typical of this mechanism. The molecular chain is not cleaved, but there is a significant decrease in molecular weight as volatile fragments are eliminated. Examples of polymer pyrolysis will be discussed in detail in subsequent sections.

1) Pyrolysis Mechanisms

In his early experiments on rubber pyrolysis, Williams obtained only a small amount of the volatile monomer, isoprene. The bulk of the reaction product was an intractable, tarry mass. During pyrolysis, natural rubber degrades only partially by depolymerization, random chain scission being the predominant reaction. Other polymers, e.g. poly(methyl methacrylate), pyrolyze almost completely by depolymerization. The relative contribution of chain scission versus depolymerization depends on polymer structure as shown in Table B-2.

It is apparent from the data on polystyrene and its related polymers that the number and size of groups attached to the alpha carbon atom play an important role. In polystyrene, this carbon has an aromatic ring as one substituent and the other substituent is hydrogen. Monomer accounts for less than half of the degradation products from this polymer. Replacement of hydrogen by the slightly larger deuterium atom increases monomer yield significantly, and when the second substituent is a methyl group, depolymerization is almost complete. Monomer yield during pyrolysis is also dependent on temperature but to a lesser extent.

Poly(methyl methacrylate) (PMMA) is a clear, transparent polymer often used as a glazing material. However, this polymer depolymerizes almost completely when

Table B-2. Monomer yield in the pyrolysis of selected polymers

Polymer	Structure	Monomer yield (%)
Polyoxymethylene	$-- \{CH_2-O\} ---$	100
Polytetrafluoroethylene	$-- \{CF_2-CF_2\} ---$	96
Poly(methyl methacrylate)	$---\begin{bmatrix} CH_3O-C=O \\ \vert \\ -CH_2-C- \\ \vert \\ CH_3 \end{bmatrix}---$	95
Poly(methyl acrylate)	$---\begin{bmatrix} CH_3O-C=O \\ \vert \\ -CH_2-CH- \end{bmatrix}---$	1
Polystyrene	$---\begin{bmatrix} C_6H_5 \\ \vert \\ CH_2-CH- \end{bmatrix}---$	41
Poly(α-deutero styrene)	$---\begin{bmatrix} C_6H_5 \\ \vert \\ CH_2-CD- \end{bmatrix}---$	68
Poly(α-methyl styrene)	$---\begin{bmatrix} C_6H_5 \\ \vert \\ CH_2-C- \\ \vert \\ CH_3 \end{bmatrix}---$	100
Polyisoprene	$---\begin{bmatrix} CH_3 \\ \vert \\ CH_2-C=CH-CH_2 \end{bmatrix}---$	11
Polyethylene	$-- \{CH_2-CH_2\} ---$	1
Poly(vinyl chloride)	$-- \{CH_2-CHCl\} ---$	0

heated above 300 °C. A similar polymer, poly(methyl acrylate) is more resistant to depolymerization. Copolymers of methyl methacrylate and methyl acrylate have been prepared as possible replacements for PMMA. These copolymers would be expected to be more resistant to depolymerization than the homopolymer. The anticipated improvement in stability was realized, but the copolymers did not have properties making them suitable as replacements for PMMA.

In addition to random chain scission and depolymerization, there is a third general degradation mechanism, best exemplified by the pyrolysis of poly(vinyl chloride) (PVC). This polymer loses hydrogen chloride sequentially along its backbone chain. When approximately nine conjugated double bond sequences are formed, the resulting polyene is an intense chromophore which accounts for the rapid discoloration of the polymer. Details of this reaction and other mechanisms of pyrolysis will be reviewed in subsequent discussions on the thermal degradation of specific polymers.

For many years there has persisted an hypothesis that polymer molecules contain weak links at which degradation is initiated. It is now generally agreed that many polymers contain such irregularities in their molecular structure. If bonds in these abnormal structures have lower dissociation energies than other bonds in the polymer, degradation might well be expected to start at these points. The head-to-head configuration, which can occur as an imperfection in PVC would be a typical weak link at which degradation might be initiated. There are, however, other irregularities in this polymer which could be primary sites for initiation.

Recently, data [7] have been reported supporting the existence of weak links in radical-initiated polystyrene when pyrolyzed at low temperatures. Though these apparent weak bonds, believed to be randomly dispersed, have not been identified, evidence for their existence is persuasive. The possible role of weak links will be discussed in greater detail as it relates to the thermal degradation of those polymers where there is developing evidence for their existence.

2) Pyrolysis Reactions of Selected Polymers

In the following sections, pyrolysis of a few individual polymers is discussed in some detail. Those selected are typical examples of polymers that pyrolyze primarily by one of the previously discussed mechanisms. However, pyrolysis seldom proceeds entirely by a single mechanism. The process is always complex, usually involving more than one basic mechanism or at least variations within the primary mechanism. Furthermore, it is extremely difficult to study pyrolysis reactions in the complete absence of chemical degradation. Although pyrolysis of each of the polymers to be discussed is subject to these complexities, their degradation reactions are relatively simple and have been studied extensively.

a) Pyrolysis of Poly(vinyl chloride)

In the absence of oxygen, poly(vinyl chloride) (PVC) degrades thermally by sequential elimination of hydrogen chloride (HCl) which leads to discoloration. Despite the efforts of countless investigators over many years, however, considerably controversy persists over several important details of the mechanism. Both ionic and radical mechanisms have been proposed and more recently a molecular or concerted mechanism has also been suggested. Degradation by combinations of these mechanisms is now finding considerable support.

Much of the controversy over the mechanism has centered around the initiation step with the weak-link theory being an important consideration. Since PVC is much more susceptible to dehydrochlorination than are its low-molecular-weight analogues, the presence of occasional structural irregularities in the larger molecules is considered by many to be an important factor. The inverse relationship between dehydrochlorination and molecular weight has been interpreted as an indication that groups at or near chain ends are involved.

There is evidence for the following defects in the molecular structure of PVC:
Head-to-head Addition [8, 9]

$$
\text{---} \text{CH}_2\text{---CH---CH---CH}_2\text{---CH---CH}_2\text{---}\text{---}\text{---} .
$$

with Cl substituents on the 2nd, 3rd, and 5th carbons.

Allylic Chlorine [10]

$$
\text{---} \text{CH}_2\text{---CH}=\text{CH---CH---CH}_2\text{---}\text{---}\text{---} .
$$

with a Cl substituent on the 4th carbon.

Branching (tertiary chlorine) [11–13]

$$
\begin{array}{ccccccc}
 & Cl & & Cl & & Cl & \\
 & | & & | & & | & \\
---CH_2-C-CH_2-CH-CH_2-CH---\,. \\
 & | & & & & & \\
 & CH_2-CH--- & & & & & \\
 & | & & & & & \\
 & Cl & & & & &
\end{array}
$$

Chlorine atoms at each of these defect points could be sites for initiation of dehydrochlorination. Catalyst residues at chain ends may also contribute to initiation. There is recent evidence [14] indicating that normal groups in the backbone chain could also be involved under certain reaction conditions.

The importance of defect structures on initiation of PVC dehydrochlorination has been confirmed recently by Starnes and coworkers [15] who have succeeded in replacing labile chlorine atoms with stable ligands by the following reaction:

$$
\begin{array}{c}
Cl \\
| \\
---CH_2-C-CH{=}CH---- + ML_2 \rightarrow \\
| \\
H \qquad\qquad (MLCl)
\end{array}
$$

$$
\begin{array}{c}
L \\
| \\
---CH_2-CH-CH{=}CH---- + MLCl \\
(MLCl_2)
\end{array}
$$

in which M is Ba, Cd, Zn, Ca or Pb, and the ligand, L, is —O_2CR, —OR, —SR, etc. When the ligand is —SR, stability of the modified PVC has been shown to be directly proportional to the amount of chemically-bound sulfur. Attempts to identify the labile group(s) that are replaced have not as yet been conclusive. It is quite likely that more than one defect group is responsible for initiating dehydrochlorination, depending on composition of the particular lot of polymer and on the conditions under which degradation takes place.

There is evidence that dehydrochlorination of PVC may proceed by a radical, ionic or concerted mechanism. This latter reaction path is a modification of the ionic mechanism in which HCl is eliminated in a one-step process involving highly-ionized chlorine. Evidence for the radical mechanism is based on the inhibition of degradation by conventional radical traps [16], by studies of decomposition energetics [17], and from the degradation behavior of mixtures of PVC with polystyrene [18] or with polypropylene [16]. The effect of solvents on the reaction [19] and the decomposition of model allylic chlorides in solution [20] are believed to support the ionic mechanism. A number of theoretical studies suggest the possibility of an ionic mechanism while others [19] appear to support the concerted, one-step elimination of HCl.

The role of liberated HCl on further dehydrochlorination has added to the confusion in resolving the mechanism. Evidence [20] has been presented to show that HCl inhibits degradation. This conclusion can be explained on the basis of addition to the double bonds that form, thus reducing the length of conjugated sequences responsible for color formation. This effect is observed, however, only under certain experimental conditions. It is now generally agreed that liberated HCl acts as a strong catalyst for dehydrochlorination, primarily involving chloroallylic groups in the chain [22].

Although considerable progress has been made in elucidating the mechanism for PVC pyrolysis, several important points remain to be resolved. The consensus at this time is that concurrent radical and ionic mechanisms are involved, and that one or more structural defects are the sites for initiation. Several excellent reviews [13, 14, 23, 24] have appeared in the past decade, and the reader is referred to these for a more complete discussion of this extremely complex mechanism.

b) Pyrolysis of Polyoxymethylene

Polyoxymethylene (POM), a polymer of formaldehyde, belongs to the generic class of polyethers or acetal resins. Its rapid depolymerization in the absence of oxygen limited the development of this polymer for many years. Formaldehyde reacts spontaneously to form low-molecular-weight oligomers which rapidly revert back to monomer. Reaction is initiated at chain ends and depolymerization then proceeds sequentially along the backbone chain. Stable polymers of formaldehyde were not known until MacDonald [25] developed a process for endcapping the macromolecules through acetylation of the terminal hydroxyl groups,

$$----CH_2-O-CH_2-O-CH_2-OH$$
$$+ (CH_3CO_2)_2O \rightarrow ----CH_2-O-CH_2-O-CH_2-OCOCH_3.$$

This and similar reactions have made possible the polymerization of formaldehyde into high-molecular-weight polymers having considerable commercial importance. However, end-capped POM is still vulnerable to random cleavage along the backbone chain. This reaction would then be followed by depolymerization proceeding toward the original chain ends.

$$---CH_2-O-CH_2-O(CH_2-O)_nCH_2-OCOCH_3$$
$$\downarrow$$
$$----CH_2-O^\bullet + {}^\bullet CH_2-O-(CH_2-O-)_n-CH_2-OCOCH_3.$$
$$\downarrow$$
$$n+1\ CH_2O$$

Recent studies [26] have shown that commercial POM contains both aldehydic and α,β-unsaturated carbonyl groups.

c) Pyrolysis of Polypropylene

Thermal degradation of polypropylene (PP) occurs primarily by random scission along the backbone chain with formation of large radical fragments as the initial step and with little or no initial weight loss. Subsequent reactions of intramolecular chain transfer then take place leading to the formation of many, low-molecular-weight hydrocarbon products (Table B-3). This mechanism is in contrast to the elimination mechanisms just discussed in which either substituent groups are split off as in PVC or monomer units are eliminated sequentially as in POM. Random chain scission is the dominant mechanism in the pyrolysis of polyolefins and most vinyl polymers.

Many polymers degrade thermally by combinations of the basic mechanisms. Polystyrene, for example, degrades by random chain scission in the temperature range 280–

Table B-3. Volatile products from the pyrolysis of polypropylene at 438 °C [31]

Product	Yield in weight percent from	
	Atactic PP	Isotactic PP
Methane	0.1	0.09
Ethane	1.2	1.4
Propylene	12.2	10.7
Isobutylene	3.2	2.5
2-Pentene	15.8	15.5
3-Methyl-1-pentene	11.4	10.2
3-Methyl-3,5-hexadiene	2.0	1.7
4-Methyl-3-heptene [a]		
2,4-Dimethyl-heptadiene [a]		
2,4-Dimethyl-heptene	39.3	42.6
$C_{10}H_{20}$ [b]	2.4	2.1
4,6-Dimethyl-3-nonene	1.3	1.6
2,4,6-Trimethyl-8-nonene	9.5	10.2
$C_{13}H_{24}$ [b]	1.7	1.6

[a] Incompletely resolved
[b] Identity not reported

300 °C [27] with negligible formation of volatiles. At higher temperatures, however, elimination reactions take place at a significant rate with yields of the monomer as high as 40 weight percent observed at elevated temperatures [28]. PP degrades by both scission and elimination mechanisms, but its pyrolysis in the absence of oxygen yields much smaller amounts of monomer, chain scission being the dominant reaction even at elevated temperatures.

Most of the early studies [29, 30] on the pyrolysis of PP were conducted under high vacuum. More recently, Chien and coworkers [31, 32] have investigated the thermal degradation of this polymer in a stream of inert gas, e.g. helium. When primary products of decomposition are swept out of the reaction zone, secondary reactions are minimized. This latter technique is considered to be more realistic in studying the burning of polymers in which pyrolysis within the polymer bulk is the initial step, and volatiles are swept away from the area where pyrolysis is taking place. Of course, different secondary reactions occur in the flame zone above the burning polymer.

There are differences, largely quantitative, between the products formed during the thermal degradation of PP in vacuum and in an inert gas stream. However, reactions leading to the products listed in Table B-3 are similar and follow the mechanism under helium as proposed by Chien [31, 32].

Initiation:

$$
\begin{array}{c}
\quad CH_3 \quad\quad CH_3 \quad\quad CH_3 \quad\quad CH_3 \\
\quad | \quad\quad\quad | \quad\quad\quad | \quad\quad\quad | \\
---CH_2-CH-CH_2-CH-CH_2-CH-CH_2-CH--- \longrightarrow
\end{array}
$$

$$
\begin{array}{c}
CH_3 \quad\quad CH_3 \quad\quad\quad CH_3 \quad\quad CH_3 \\
| \quad\quad\quad | \quad\quad\quad\quad | \quad\quad\quad | \\
---CH_2-CH-CH_2-CH^{\bullet} \; + \; {}^{\bullet}CH_2-CH-CH_2-CH---
\end{array}
$$

I

Weak links are believed to be primary sites for initiation. The secondary radical (I) then undergoes intramolecular chain transfer reactions as follows.

Chain Transfer:

$$
\begin{array}{c}
CH_3-\overset{\bullet}{C}H \quad\overset{\displaystyle CH_2}{\diagdown}\quad CH-CH_3 \\
\mid \qquad\qquad\qquad\qquad \\
CH_2-\underset{H}{C}-CH_2-CH-CH_2-CH-- \\
\overset{CH_3}{\mid}\qquad\overset{CH_3}{\mid}\qquad\overset{CH_3}{\mid}
\end{array}
\longrightarrow
$$

I

$$
\begin{array}{c}
CH_3-CH \quad\overset{\displaystyle CH_2}{\diagdown}\quad \underset{H}{CH}-CH_3 \\
\mid \\
CH_2-\underset{CH_3}{C}=CH_2
\end{array}
\;+\; {}^{\bullet}\overset{CH_3}{CH}-CH_2-\overset{CH_3}{CH}-CH_2--
$$

Formation of Saturated Hydrocarbons:

$$
---\overset{CH_3}{CH}-CH_2-\overset{CH_3}{CH}-CH_2-\overset{CH_3}{CH}-CH_3
$$

$$
\rightarrow\; ---\overset{CH_3}{CH}-CH_2-\overset{CH_3}{CH}-CH_2^{\bullet} \;+\; {}^{\bullet}\overset{CH_3}{CH}-CH_3 \,.
$$

This scission at the chain end is then followed by hydrogen abstraction to form propane and another polypropylene radical (II).

$$
---\overset{CH_3}{CH}-CH_2-\overset{CH_3}{CH}-CH_2---\;+\;{}^{\bullet}\overset{CH_3}{CH}-CH_3 \xrightarrow{PP} H\overset{CH_3}{CH}-CH_3 \;+\; PP^{\bullet}
$$

(PP)

II

Other saturated products are presumed to be formed by a similar mechanism.

Depolymerization:

$$
---CH_2-\overset{CH_3}{CH}-CH_2-\overset{CH_3}{CH}{}^{\bullet} \;\rightarrow\; ---CH_2-\overset{CH_3}{CH}{}^{\bullet} \;+\; CH_2=\overset{CH_3}{CH} \;.
$$

Composition of the volatile fraction is somewhat different as obtained from atactic or isotactic PP, and also varies with the temperature and time of heating.

Though not as firmly established, similar reactions are believed to take place in other polymers which degrade by chain scission. In all of these reactions, volatiles accumulate slowly via a series of secondary reactions following the initial chain scission. Those polymers which pyrolyze primarily by elimination mechanisms, on the other hand, yield large volumes of volatiles in the initial stage of degradation. Since only a few molecules react initially, there is little change in the molecular weight of the sample until reaction has proceeded extensively.

d) Pyrolysis of Copolymers and Polymer Blends

The use of copolymers and polymer blends is increasing rapidly as various applications require properties not found in single polymers. Thermal degradation of these materials presents some unique features in addition to the basic mechanisms discussed so far. Impact modified polymers and heat-resistant copolymers are typical examples. Many important examples of both blends and copolymers have been studied [33, 34] in recent years.

The blend of polystyrene and polyisoprene is an example of an impact-modified polymer. In the pyrolysis of this blend [35], each component degrades in a manner different from that observed when polyisoprene is degraded separately. Thermal gravimetric analysis shows that both the rate of volatile formation and the rate of chain scission of polyisoprene are reduced when blended with polystyrene, but products of the pyrolysis are essentially identical, qualitatively and quantitatively. Likewise, polystyrene appears to be stabilized against thermal degradation at 340 °C when blended with polyisoprene although its chain scission appears to be accelerated at 292 °C.

These data can be explained by assuming that polyisoprene generates small radicals as it undergoes chain scission in the first stage of degradation. As these radicals diffuse into the polystyrene phase, hydrogen abstraction takes place, and the radicals which would normally contribute to degradation of the polyisoprene are thus stabilized. Subsequently, polystyrene chains undergo scission at sites adjacent to the radical centers. The resulting polystyrene radicals decompose only slowly below 300 °C. Above this temperature, there is an apparent inhibition of its depolymerization by dipentene formed from the degrading polyisoprene. This inhibition could occur through hydro-

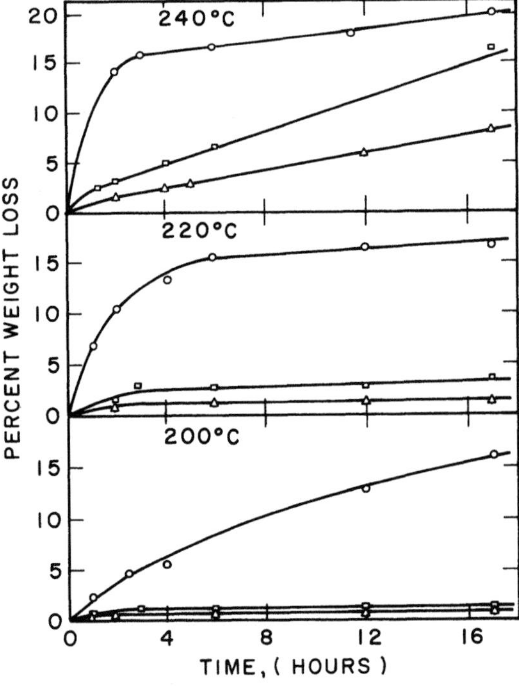

Fig. B-1. Time dependence of chain scissions during thermal degradation at 200–240 °C. o, PMMA; □, 50/1 copolymer; △, 20/1 copolymer (note the different weight loss scales). (Reprinted with permission of Applied Science Publishers, Ltd.)

gen abstraction by polystyrene radicals from the dipentene or by coupling of the macroradicals with allylic radicals.

Poly(methyl methacrylate) (PMMA) depolymerizes almost completely when heated in the absence of oxygen. Commercial applications of PMMA may require improved heat resistance and one route to achieve this is through copolymerization. The thermal degradation of copolymers of methyl methacrylate and maleic anhydride have been investigated by Grassie and Davidson [36]. The extent of thermal stabilization obtained with varying amounts of the comonomer at three different temperatures is shown in Fig. B-1. Clearly the presence of the comonomer inhibits depolymerization of PMMA. The effect is explained on the basis that depolymerization of PMMA is initiated at unsaturated end groups. Weak links have also been proposed as initiation sites. Thermogravimetric analysis suggests that although maleic anhydride accelerates the chain scission of PMMA, depolymerization is inhibited. As depolymerization proceeds from the point of chain scission, monomer units of methyl methacrylate split off until the first maleic anhydride unit is reached. Depolymerization stops at this point.

II) Oxidative Degradation – An Historical Introduction

The oxidative degradation of polymers is a general phenomenon studied extensively by many investigators over several decades. Failure of polymers by this mechanism is of considerable commercial importance. The complex series of reactions responsible for oxidative degradation will first be reviewed historically and then followed by more detailed discussion of recent developments in the oxidation of a few selected polymers.

Irreversible reaction with oxygen is the basis for the most common degradation of polymers. Without exception, all commercially-important polymers undergo reactions with oxygen, leading eventually to changes in molecular structure [37] or in morphology [38]. As a result of these reactions, critical mechanical, dielectric or aesthetic properties may change beyond acceptable limits. In addition to the ubiquitous molecular form of oxygen, it has been suggested that ozone and perhaps singlet oxygen also contribute to polymer degradation.

The initial effect of oxidative degradation is often quite subtle and difficult to detect. Chemical changes occur at random sites in an infinitesimal number of molecules in the mass. These initial reactions are so infrequent that detection eludes even the most sensitive analytical techniques. However these incipient reactive sites then initiate further oxidation during normal service life until obvious changes are evident. Thus it is important to recognize each stage in a polymer's life cycle at which oxidation could take place.

Oxidative degradation can occur during the synthesis of highly sensitive polymers. Once removed from the synthetic environment, sensitive polymers oxidize slowly even at ambient temperatures – unless adequately protected. All polymers are vulnerable to oxidation during thermoforming. The environment within an extruder or an injection molding machine contains sufficient oxygen to initiate degradation. Since fabrication is usually carried out well above the polymer's T_m, initiation sites for future oxidation are likely to be formed. Even in those processes which essentially exclude oxygen, enough of the reactant is brought in with the polymer to initiate oxidation. Recognition of the potential for degradation in these early stages has lead to development of special stabilizers designed to protect during thermoforming. Every exposure to heat, as a polymer is compounded or pellitized, presents ample opportunity for incipient oxidation to occur.

Once the polymer has been fabricated into a finished product, oxidation will proceed, though more slowly, throughout the service life. Failure is inevitable when the polymer is exposed for a long enough period. When long service life is required, almost every commercial polymer must be stabilized to inhibit oxidative degradation. Recently, attention has been directed to the importance of protecting polymer scrap which is to be recycled. Unless more protection was provided than needed for the original application, oxidation could damage the scrap as it is exposed to additional heat treatment during recycling.

Most of the oxidative degradation of a polymer occurs during long-term aging, and a different type of stabilizer is required to protect the material under these conditions. There is perhaps a connection between the sites of incipient oxidation and the weak links which are believed to initiate thermal degradation in the absence of oxygen. Oxidation proceeds under service conditions as the polymer absorbs energy from one or more sources. Thermal and radiation energy (ultraviolet and high-energy radiation) are the major factors responsible for accelerating oxidative degradation, but mechanical energy can also contribute to these reactions. Each of the major energy sources promotes oxidation by a somewhat different mechanism, and appropriate stabilizers must be used for each exposure condition. Frequently, a combination of energy sources is involved in the degradation of a polymer.

1) Thermal Oxidation

The combination of oxygen as the reactant and heat as the energy source is a major factor in polymer degradation [39]. Even the most inherently-stable polymers succumb to this combination when exposed to sufficiently high temperatures for long periods. It is not surprising therefore that thermal oxidation of polymers has been studied so extensively. Emphasis has been placed on mild conditions of thermal exposure, usually referred to as autoxidation, the oxidation which occurs between ambient temperature and about 200 °C. The autoxidation of hydrocarbon polymers is now well understood as a result of the extensive research conducted at the Natural Rubber Producer's Research Association [40, 41]. Based on this early research on model compounds for rubber, Shelton [42] has proposed the following kinetic scheme for the autoxidation of hydrocarbon polymers,

Initiation:

$$ROOH \rightarrow RO^{\cdot} + HO^{\cdot}$$

or

$$2ROOH \rightarrow RO^{\cdot} + ROO^{\cdot} + H_2O.$$

Propagation:

$$ROO^{\cdot} + RH \rightarrow ROOH + R^{\cdot}$$

$$R^{\cdot} + O_2 \rightarrow ROO^{\cdot}.$$

14

Oxidative Chain Branching:

$$ROOH \rightarrow RO^{\bullet} + HO^{\bullet}$$
$$RO^{\bullet} + RH \rightarrow ROH + R^{\bullet}.$$

Eventually autotermination would take place either by coupling or disproportionation of propagating radicals, but from the practical point of view, the induced termination reactions of stabilization are of much greater, commercial importance. In this reaction scheme, RH could represent either a low-molecular-weight hydrocarbon or a hydrocarbon polymer.

Initiation of the oxidation of pure low-molecular-weight hydrocarbons is assumed to occur by attack of free radicals, formed during hydroperoxide homolysis. Minute traces of these hydroperoxides could be present initially. Initiation of polymer autoxidation is probably much more complex. Hydroperoxides are assumed to be the primary source of radicals which initiate oxidation in polymers, but various additives including stabilizers may be involved in the initiation step. Sensitizing groups formed during prior processing may also lead to hydroperoxide formation.

Propagation is a chain reaction in which a single initiation event could lead to further reaction in hundreds of other molecules. Chain reactions of propagation and oxidative chain branching are responsible for the rapid degradation that takes place during autoxidation. As hydroperoxides accumulate, the rate of branching increases, and this leads to the autocatalytic stage of autoxidation. This relationship between hydroperoxide concentration and autocatalysis is shown schematically in Fig. B-2. The maximum hydroperoxide concentration is reached at that point where hydroperoxide decomposition exceeds its rate of formation, and this coincides with the rapid increase in reaction rate or autocatalysis. A similar relationship has been demonstrated [43] in the autoxidation of polyethylene by using infrared spectroscopy to determine hydroperoxide concentration. It is believed that modifications of this mechanism are applicable to the autoxidation of other polymers in which there are significant hydrocarbon segments.

The chemical structure of hydrocarbons and hydrocarbon polymers plays an important role in autoxidation. This is evident in differences observed in the rates of oxidation of linear polyethylene, branched polyethylene and polypropylene. At those points in these polymers where branching occurs in the backbone chain, a hydrogen is attached to a tertiary carbon atom.

$$
\begin{array}{ccc}
 & \begin{matrix} CH_3 \\ | \\ (CH_2)_3 \\ | \end{matrix} & \begin{matrix} CH_3 \\ | \end{matrix} \\
---[CH_2-CH_2]_n--- & ---[CH_2-CH_2]_n-CH--- & ---[CH-CH_2]_n--- \\
\text{Linear} & \text{Branched} & \text{Polypropylene} \\
\text{Polyethylene} & \text{Polyethylene} &
\end{array}
$$

——— Susceptibility to Oxidation ———▶

The bond between carbon and the hydrogens at these branch points has a lower dissociation energy than that between hydrogens and carbons in the methylene groups along the chain. These labile hydrogens are likely points for initiation of autoxidation

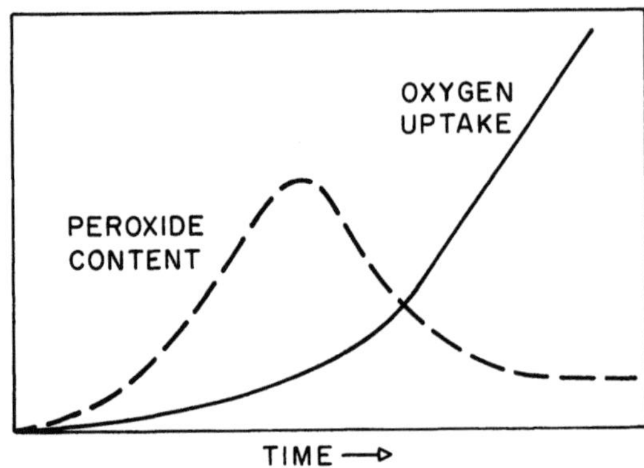

Fig. B-2. Relation between oxidation rate and hydroperoxide accumulation. (Reprinted with permission of The Society of Plastics Engineers)

and as a result branched polyethylene is measurably less stable to oxidation than the linear modification. Polypropylene, with a methyl branch at every other carbon along the backbone chain, is less stable to autoxidation than either of the polyethylenes. The extent to which hydrogen at branch sites in low-density polyethylene contributes to oxidation, however, decreases at higher temperatures. Under these conditions, oxidation proceeds too rapidly for selective hydrogen attack to occur. These differences in resistance to autoxidation are reflected in the amounts of stabilizers required to provide protection to these polymers.

The autoxidation of polystyrene presents an anomoly which would seem to refute the above conclusion. Although this polymer also has a labile hydrogen at alternating carbons along the backbone chain, it is quite stable to autoxidation. It has been suggested [44] that stability of this polymer results either from steric protection of the labile hydrogens by the bulky aromatic rings, or from the loss of resonance energy caused by unfavorable orientation of phenyl groups in the structure [45]. In general, however, whenever branched groups are present which lower the dissociation energy of carbon-hydrogen bonds, the polymer will have poor resistance to autoxidation.

Physical factors can also affect autoxidation rates. For example, certain semi-crystalline polymers, polyethylene being a classical example, have a crystalline structure so compact that oxygen cannot penetrate into crystallites. Oxidation of such polymers is restricted to the disordered or amorphous regions, except for slow, surface reaction of crystallites. This is evident in Fig. B-3 in which autoxidation rates and the limiting amount of oxygen reacting are shown for linear and branched polyethylenes, above and below the T_m. At a temperature above the T_m, both polymers react at about the same rate, and approximately the same amount of oxygen is absorbed before autoxidation subsides. Below the T_m, however, linear polyethylene oxidizes more slowly and absorbs significantly less oxygen than the branched modification. The relative amounts of oxygen absorbed are proportional to the amorphous content of each polymer, about ten percent for linear and about forty percent for branched polyethylene. Polymers whose crystalline structure is permeable to oxygen, absorb approximately the same amount of oxygen just above as just below their T_m.

In the crystalline regions of polyethylene, individual crystallites are joined by bundles of tie molecules which constitute the disordered region [46]. A photomicrograph of these structures is shown in Fig. B-4. The maximum level of crystallization is restricted by these bundles of tie molecules. Reaction with a small amount of oxygen can cause

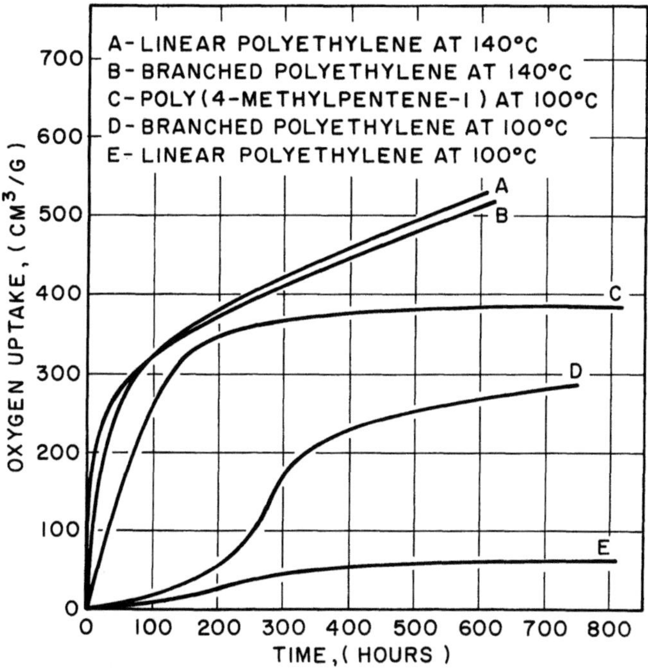

Fig. B-3. Effect of morphology on the oxygen uptake of polyolefins. (Reprinted with permission of John Wiley and Sons, Inc.)

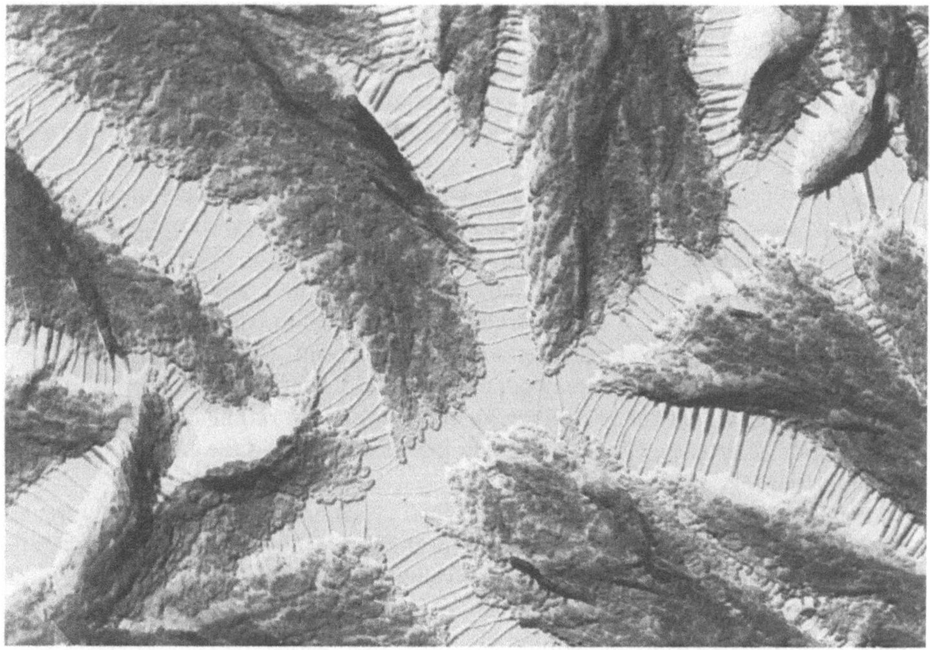

Fig. B-4. Bundles of tie molecules connecting individual crystallites. (Reprinted with permission of The American Institute of Physics)

limited scission of tie molecules and crystallization then proceeds to a slightly higher extent [38]. Such an increase in crystallinity could be considered as degradation in applications in which a critical balance between ordered and disordered regions is required for the combination of strength and flexibility. Examples of changes in polymer morphology as a result of autoxidation are not uncommon, but these may not necessarily lead to failure.

2) Photooxidation

Ultraviolet radiation is responsible for the photodegradation of polymers. As in thermal oxidation, however, significant differences exist among polymers in their resistance to photooxidation. Polytetrafluoroethylene, poly(methyl methacrylate) and the urea-formaldehyde resins have exceptional stability. Most other polymers have only moderate to poor resistance to light-induced oxidation. The polyolefins, polystyrene, poly-(vinyl chloride), polyacetals, polyisoprene, and polyacrylonitrile are among the more light-sensitive polymers. In contrast to thermal oxidation, which can occur in subsurface regions of permeable polymers, photooxidation is limited to reactions at or near the polymer surface.

a) Outdoor Weathering

Outdoor weathering of polymers is a complex process involving both thermal and photooxidation. The rapid failure of many polymers and the level of protection that can be provided are a major factor restricting outdoor applications. Since in outdoor weathering it is difficult to isolate those reactions that are photo-induced from those due to thermal effects, basic mechanisms for true photooxidation are studied more effectively under indoor, laboratory conditions. Nonetheless, outdoor weathering and the losses which result are of considerable commercial importance. It is therefore appropriate to consider briefly the basic principles that influence polymer degradation in an outdoor environment.

Radiation from the electromagnetic spectrum of the sun that reaches the earth's surface varies seasonally and with geographic location. The solar spectrum extends from the ultraviolet, through the visible and into the infrared region. Thermal degradation results from absorption of infrared radiation whereas the ultraviolet (uv) component between 290 and 400 nm, often referred to as the near-ultraviolet, is responsible for photooxidation. Ultraviolet radiation emitted by the sun is filtered out below about 290 nm by the ozone layer in the upper atmosphere. Variations in the level of uv radiation with location are due to differences in ozone concentration and the intensity of sky radiation. Scattering of solar radiation also contributes to the amount of uv radiation reaching the earth's surface.

Some polymers absorb uv radiation through groups in their normal structure, but quite frequently it is the presence of structural irregularities or associated impurities that are the primary uv absorbers. Polyethylene, in its ideal structure, should contain no groups capable of absorbing in the near uv. Yet this polymer is much less stable than would be anticipated from the stability of its low-molecular-weight analogues. It is now recognized that the instability of polyethylene is due to the presence of traces of carbonyl and/or hydroperoxy groups, presumably formed during processing. Certain

Table B-4. Activation spectra maxima for various polymers

Polymer	Activation Spectra Maxima
Polyesters	325 nm
Polystyrene	318 nm
Polyethylene	300 nm
Polypropylene (unstabilized)	310 nm
Poly(vinyl chloride)	320 nm
Poly(vinyl acetate)	280 nm
Polycarbonate	295 nm
Poly(methyl methacrylate)	290–315 nm
Polyoxymethylene	300–320 nm
Cellulose Acetate Butyrate	295–298 nm

(Data in part from Hirt and Coworkers) [47]

catalyst residues attached to polymer molecules may also function as sites for uv absorption.

Individual polymers absorb uv radiation within specific wave length regions, exhibiting activation spectra maxima at which each is most vulnerable to photooxidation [47]. Typical maxima for a few representative polymers are shown in Table B-4. Selection of uv stabilizers should take into account these maxima in order to provide the greatest level of protection. Many uv absorbers have been developed which strongly absorb radiation at damaging frequencies.

Photooxidation produces a variety of physical and chemical changes. Discoloration and surface cracking are visible evidence of degradation. The adverse effect on mechanical and dielectric properties can be detected instrumentally. Chain scission and crosslinking are the general reactions that take place, often accompanied by formation of oxygen-containing groups.

In the outdoor weathering of polymers, absorption of infrared radiation and the accompanying temperature rise can increase the rate of photochemical reactions. Conversely, products from polymer photolysis may serve as initiators or chain carriers for thermal oxidation. Because of these and other complications, mechanistic studies are usually done in the laboratory where thermal effects can be minimized. As in thermal oxidation, model compounds have been used extensively in these studies.

b) Mechanisms for Photooxidation

It is generally accepted that reactions involved in photooxidation are similar to those of thermal oxidation. Both are now believed to proceed by a free-radical, chain mechanism, but there are important differences between them. For instance, the length of oxidative chains in the propagation phase is much shorter than in thermal oxidation, and initiation reactions are more complex. Also, photooxidation is primarily a surface reaction, suggesting that the effect of uv radiation may not extend into the polymer bulk to a significant extent. Permeation of oxygen into the polymer bulk could also restrict photooxidation to surface layers.

Some uncertainty still persists over the initiation reaction in photooxidation, and this is not surprising since it is extremely difficult to precisely identify the very first reaction products or those sites in polymer molecules where initial reactions occur. As in thermal oxidation, it has been necessary to use model compounds in order to investigate individual chemical groupings believed to exist in complex polymer molecules. Small molecules can also be purified much more completely than polymers, thus eliminating side reactions.

Early investigators were puzzled by the rapid photooxidation of the polyolefins as contrasted to the stability of simple paraffins. If polyethylene consisted only of a carbon-carbon backbone chain with nothing but hydrogen atoms as its substituents, this polymer should have negligible absorption in the uv region, and should be as photostable as the simple paraffins. Double bond structures are usually responsible for absorption of uv radiation, and polymers invariably contain such groups.

The presence of carbonyl groups in polyethylene was suggested by Pross and Black[48] in 1950, and Burgess[49] proposed a mechanism for photooxidation based on ketone scission reactions which are termed Norrish Type I and Type II reactions.

Norrish Type I:

$$---CH_2-CH_2-\overset{\overset{\displaystyle O}{\|}}{C}-CH_2-CH_2----$$

$$\xrightarrow{h\nu} ---CH_2-CH_2\overset{\overset{\displaystyle O}{\|}}{C}{}^{\bullet} + {}^{\bullet}CH_2-CH_2---.$$

$$\downarrow$$

$$---CH_2-CH_2^{\bullet} + CO$$

Norrish Type II:

$$---CH_2-CH_2-\overset{\overset{\displaystyle O}{\|}}{C}-CH_2-CH_2-CH_2----$$

$$\xrightarrow{h\nu} ---CH_2-CH_2-\overset{\overset{\displaystyle O}{\|}}{C}-CH_3 + CH_2{=}CH-CH_2---.$$

In those structures where a hydrogen is present at the gamma carbon, quantum yield studies indicate that the Type II reaction is favored[50]. Though both reactions result in chain scission, it is important to note that only the Type I reaction yields free radicals. This mechanism suggests[51] that in the photooxidation of polyethylene, carbonyl groups are the major sites of reaction as well as the groups primarily responsible for absorption of uv radiation.

Carlson and Wiles[52] in their studies of polypropylene photooxidation have emphasized the important role played by hydroperoxides. Initially, these reactive intermediates were thought to play only a minor role in the photooxidation of polyolefins since they are present in only small amounts and no appreciable concentrations are detected until the reaction is well underway. However, hydroperoxides also absorb uv radiation and decomposition of these reactive intermediates into radicals must be taken into account.

Recently, Guillet[53] in reviewing the data responsible for conflicting theories on the photooxidation of polyolefins concluded that degradation is initiated by free radicals, formed after absorption of a quantum of uv energy, either by hydroperoxide or carbonyl groups initially present. Reaction then continues by a radical-initiated, chain mechanism. Careful analysis of all available data indicates that radicals formed by the photolytic decomposition of hydroperoxides are the propagating species in photooxidation. This would account for the small amounts of hydroperoxides found in oxidizing polyolefins, and is in accord with the recognized rapid photolysis of these reactive intermediates.

Carbonyl groups accumulate rapidly as photooxidation proceeds, and these intermediates absorb uv radiation almost four times as efficiently as hydroperoxides. Thus carbonyl groups appear to be the primary energy absorbers. Furthermore, they are considered to be sensitizers for the induced decomposition of hydroperoxides. Although other mechanisms for hydroperoxide sensitization have been suggested, this secondary role of carbonyl groups supports a free-radical, chain mechanism similar to that of thermal oxidation. The rapid decomposition of hydroperoxides, sensitized by carbonyl groups or by some other mechanism, explains their low concentration in oxidizing polymers. Oxidative chains would then be expected to be short since the radicals formed rapidly become involved in the initiation of new oxidative chains. The increasing concentration of radicals from hydroperoxide photolysis would also overwhelm those stabilizers which are effective in inhibiting thermal oxidation. At one time, failure of these stabilizers to protect against photooxidation was interpreted as evidence that this degradation mechanism did not proceed by a chain reaction. The quantum yield for chain scission of carbonyl compounds is only 2 to 3 percent of that for hydroperoxides, but both groups probably contribute to chain scission.

Although details of this mechanism have not as yet been completely supported by experimental evidence, it seems to fit most data available on the photooxidation of polyolefins. Many complicating factors enter into photooxidative reactions of commercial polymers, however, as a result of impurities and structural irregularities that are usually present. Singlet oxygen and ozone have been suggested [51] as possible initiators of photooxidation. Traces of residual catalyst or added stabilizers may influence the mechanism as proposed for unmodified polyolefins. The photodegradation of other polymers has not been investigated as extensively as that of the polyolefins.

Developments over the past decade in the photooxidation of selected polymers are reviewed in Section III.

3) Miscellaneous Degradation Reactions

Thermal oxidation, specifically autoxidation and photooxidation as a component of outdoor weathering are the two most general types of degradation affecting all polymers. In addition to these, there are several other types of degradation which take place under unusual conditions, or which affect only a limited number of polymers. Since these degradation reactions have a lesser impact on general applications for polymers or have been encountered only recently, they have not been investigated to the same extent as thermal or photooxidation. However, some of these miscellaneous reactions are of sufficient importance to warrant a brief review with highlights on recent developments.

It has already been mentioned that high-energy radiation contributes to polymer degradation when polymers are exposed to intense radiation from natural or artificial sources. The increasing use of polymers in extraterrestrial applications and exposure near nuclear reactors are representative of these two environments. Initiation reactions, which occur under both natural and artificial conditions, are essentially the same, but in extraterrestrial applications, oxidation following initial bond scission is of course negligible. Protection against high-energy radiation reactions will be discussed under the general subject of stabilization against degradation by radiation sources.

When polymers are exposed to high-energy radiation, the initial reaction is a homolytic scission of bonds to form free radicals. When the ruptured bonds are those between substituent groups on the backbone chain, small molecular fragments are formed and if these diffuse away from the macroradical, recombination is not possible. When carbon-to-carbon bonds along the backbone chain are cleaved, however, the fragments are large radicals which are held in close proximity by a cage effect caused by the surrounding polymer mass. There is then a high probability for recombination. However, if oxygen is available, reaction with the initial radicals is likely to occur so that the backbone chain of carbon-to-carbon bonds cannot reform. This reaction is important in radiolytic reactions, but only in terrestrial applications. In space, there is little or no oxygen available for these secondary reactions.

Formation of ethylenic unsaturation during radiolysis has been attributed [54] to elimination of molecular hydrogen in a single step as represented in the radiolysis of polyethylene,

$$- - -CH_2-CH_2-CH_2-CH_2- - -$$

$$\xrightarrow{\text{Radiation}} - - -CH_2CH=CH-CH_2- - - + H_2 .$$

Formation of molecular hydrogen may also result from the following two-step reaction mechanism,

$$- - -CH_2-CH_2-CH_2-CH_2- - -$$

$$\xrightarrow{\text{Radiation}} - - -CH_2-\overset{\bullet}{C}H-CH_2-CH_2- - - + H^{\bullet}$$

$$H^{\bullet} + - - -CH_2-CH_2-CH_2-CH_2- - -$$

$$\rightarrow - - -CH_2-\overset{\bullet}{C}H-CH_2-CH_2- - - + H_2 .$$

Coupling of macroradicals leads to the formation of cross-linked polymers.

$$- - -CH_2-\underset{\bullet}{C}H-CH_2-CH_2- - -$$
$$+$$
$$- - -CH_2-\overset{\bullet}{C}H-CH_2-CH_2- - -$$

$$\rightarrow \begin{array}{c} - - -CH_2-CH-CH_2-CH_2- - - \\ | \\ - - -CH_2-CH-CH_2-CH_2- - - . \end{array}$$

Extensive cross-linking during radiation can lead eventually to gel formation. It has been suggested [55] that chain scission to produce unsaturated products occurs as follows.

$$\begin{array}{c} CH_3 \\ | \\ - - -CH_2-CH-CH_2-CH_2-CH_2- - - \end{array}$$

$$\xrightarrow{\text{Radiation}} \begin{array}{c} CH_3 \\ | \\ - - -CH_2-CH=CH_2 + {}^{\bullet}CH_2-CH_2- - - . \end{array}$$

There has been an interesting new development reported by Bovey [56] in the radiolysis of crystalline structures with a chain-folding orientation. In these crystals, molecular chains are folded with fixed spacing between folds. High-energy radiation might be expected to form cross-links if the spacing is favorable, but there is little or no flexibility in the crystallites. If cross-linking cannot take place, then scission would be the likely reaction. When crystals of the model hydrocarbon, $C_{44}H_{90}$, were exposed to gamma radiation of 53 Mrad with ^{60}Co at 25 °C, there appeared to be no chain scission. Instead gel permeation chromotography showed that about one percent of the molecules had doubled in molecular weight. Apparently radicals formed at random sites along chain folds migrated to the chain ends where coupling occurred as shown schematically below.

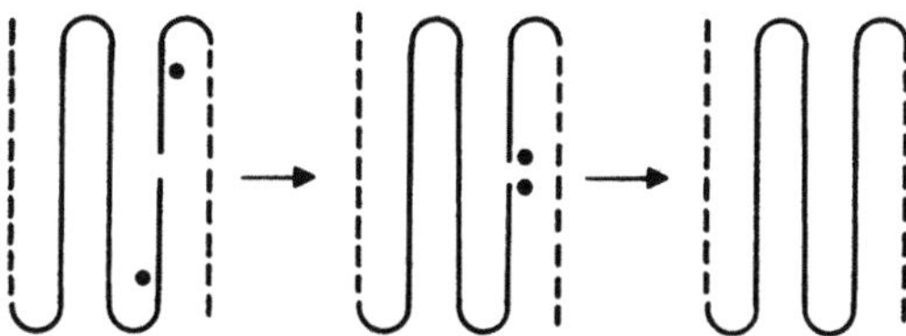

Coupling may occur either within a single fold or between adjacent layers within the crystalline structure. This reaction is of considerable importance to general concepts in polymer degradation and stabilization. In crystalline regions that are not permeable to oxygen, carbon radicals are long-lived, and apparently can migrate along the chain for significant distances. There is no oxygen within the crystallites of certain polymers, e.g. polyethylene, to form alkoxy radicals. The extent that this phenomenon takes place in semicrystalline polymers is not known, but it could explain the protection against oxidation provided by completely immobile stabilizers, e.g. carbon black. There is the possibility that termination in such cases takes place as reactive radicals migrate along or across polymer chains to reach the immobile stabilizer. This concept would also explain the efficacy of stabilizer moieties attached to a polymer chain.

Each of the radiolysis reactions discussed to this point could lead to polymer degradation, without subsequent reaction with oxygen. When oxygen is present, as is likely in all but extraterrestrial applications, the initial macromolecules undergo reactions much the same as those described in thermal oxidation. Peroxy radicals form which then abstract hydrogen from the same or other polymer molecules in a typical radical-initiated, chain mechanism.

The reaction of ozone with stressed, unsaturated elastomers has been studied extensively because of its importance in the degradation of natural rubber. This reaction is limited to those elastomers having ethylenic bonds in their backbone chain and only when they are under mechanical stress. Ozonolysis is the primary reaction in this type of degradation which is not believed to be free-radical in nature [57]. The degradation of rubber by ozone has been studied by many investigators [58–60]. Ozone has been discussed previously as it exists in the upper atmosphere, providing a filter for uv radiation below 290 nm. There is also a significant concentration of this highly reactive form of oxygen near the earth's surface, particularly in urban and industrial areas.

The effect of ozone on rubber under stress is observed as surface cracks which are always perpendicular to the direction of applied stress. When a biaxial stress is applied, the surface develops a pattern of small squares. As reaction proceeds, these cracks may propagate through the polymer, eventually causing fracture. Failure could then result from loss in mechanical strength or in surface appearance.

The basic chemical reaction involved in ozone attack is chain scission of ethylenic double bonds. Reaction is limited to the surface, probably the result of the high reactivity of ozone which decomposes on contact with most surfaces. Because of its high reactivity, it has also been suggested that ozone reacts catalytically in oxidative degradation. It has been difficult to demonstrate such a catalytic effect, however, since it is very difficult to remove the last traces of ozone from air or molecular oxygen. Other atmospheric contaminants have also been proposed as catalysts in oxidation. These include oxides of sulfur and nitrogen as well as organic peroxides emitted in automobile exhaust.

Mechanical energy in the form of applied stress promotes several types of polymer degradation in addition to its role in ozone-induced reactions. High levels of mechanical energy can cause the rupture of polymer bonds. This occurs when elastomers are masticated under high stress. Radicals formed as bonds are ruptured, and the resulting degradation causes changes in chemical structure and in mechanical properties. In the presence of oxygen, the initial radicals react rapidly to form peroxy radicals. Typical oxidative-chain reactions then follow. Mechanical stress has also been shown to accelerate oxidative degradation of polyolefins [61].

Mechanical effects have been described recently in the drawing of polypropylene monofilaments [62]. Both draw speed and low temperature, which can cause high sheer, influence the mechanical rupture of bonds during filament drawing. This draw-induced degradation is reflected in subsequent instability of the filaments to photooxidation.

Environmental stress cracking is a special type of degradation which occurs in polymers at a stress concentration lower than the polymer's ultimate strength [63]. Cracking at this critical stress occurs rapidly on contact with certain surface-active agents. These stress-cracking agents are not solvents for the polymer nor do they react with it chemically. Although the mechanism has not been established, polymer morphology is believed to be an important factor. Reaction is surface initiated, but surface cracks quickly propagate through the specimen until fracture occurs. After the sample ruptures, no chemical change can be detected at the point of fracture. Thus this mode of degradation results from purely physical changes.

Molecular weight influences environmental stress cracking, and modification in molecular weight distribution has been used effectively to reduce the susceptibility of polyethylene to this mode of failure [64]. Chemical structure is also important as evidenced by the fact that polymers vary widely in stability to environmental stress cracking. Polyolefins and polycarbonates are among the more vulnerable polymers to this mode of degradation.

Solvent crazing of stressed polymers is sometimes considered as a type of degradation. In contrast to environmental stress cracking, however, solvent crazing can be reversed if the applied stress is removed quickly [65]. The rate at which crazing occurs is also dependent on the level of applied stress [66]. When the stress is not removed quickly, solvent crazing cannot be reversed, and degradation results.

III) Recent Developments in Oxidative Degradation

Important progress has been made over the past decade in elucidating mechanisms for oxidative degradation. Selected results, summarized in the following sections, include some of these advances in thermal and photooxidation.

1) Photooxidation of Polyethylene

In thick samples of unmodified, low-density polyethylene (LDPE), the effects of photo-degradation in air are concentrated at the surface that was exposed to radiation, and decrease rapidly in going into the polymer bulk. Several investigators [67, 68] have used the concentration of carbonyl groups, measured by infrared spectroscopy, to establish the degradation profile in thick polymer samples. Under either natural or artificial weathering conditions, carbonyl concentration of LDPE falls off sharply just below the surface. Then to a depth of approximately 0.8 mm there is a continuing decrease in car-bonyl content. Carbonyl content is negligible at the center of the sample. In 25 mm thick samples, this carbonyl profile is almost identical with that measured from the side that was not exposed to uv.

Measurement [69] of the penetration of uv radiation into the polymer showed an in-tensity at the center about half of that at the exposed surface. About a third of the in-cident radiation reached the rear surface. Since carbonyl concentration falls off to a negligible level at the center where the uv intensity is still high, restriction of photo-degradation to or near the surface must be attributed to the low permeation of oxygen into the polymer. This conclusion was confirmed by exposing a similar sample with a metal film laminated to the rear surface. In this experiment, the carbonyl concentration at the rear surface was about the same as that found in the center. The metal film pre-vents penetration of air into the rear surface and photooxidation is thus restricted despite the ample intensity of uv radiation. When similar samples were exposed in pure oxygen, the deep trough in carbonyl concentration was less obvious, but oxidation at the rear surface was as extensive as in the sample radiated in air. At the center, carbonyl content was about five times as high as in the presence of air. Migration of air into LDPE is about equivalent to that of oxygen, but the amount of reactant is only about one fifth.

When 2 to 3 percent of carbon black is dispersed into LDPE, the depth of photooxi-dation is controlled by the depth of uv penetration rather than by oxygen concentration within the polymer. This capability of carbon black to screen out uv radiation accounts for the excellent protection given to most polymers by this pigment.

2) Photooxidation of Polypropylene

Recent studies [70, 71] on the photooxidation of polypropylene (PP) have helped to clari-fy the initiation step in the degradation of hydrocarbon polymers. Photooxidation of commercial samples is probably initiated by minute traces of hydroperoxide or car-bonyl groups. These impurities are believed to be formed by thermal oxidation during

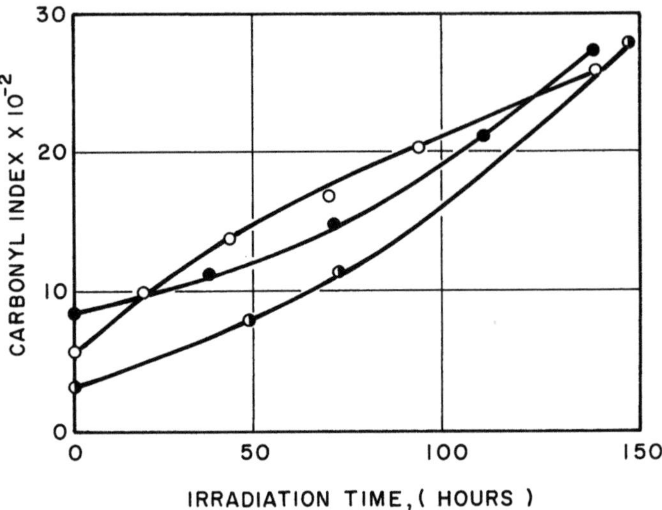

Fig. B-5. Rates of photo-oxidation of polypropylene films in a microscal apparatus after 3 h heating at 130 °C and ○: no pre-irradiation treatment, ●: pre-irradiation under nitrogen for 50 h with wavelengths > 300 nm and ◐: pre-irradiated under nitrogen for 50 h with wavelengths > 370 nm. (Reprinted with permission of Applied Polymer Science)

processing. Both groups function as sites for absorption of uv radiation, a fact which probably accounts for the poor stability of hydrocarbon polymers as contrasted to their purified, low-molecular-weight analogues. Initiation in the absence of these active groups results from a different sequence of reactions.

Allen and Fatinikun [71] subjected mildly-preoxidized polypropylene to photolysis in an inert atmosphere to destroy the photo-active groups. Pre-irradiation under nitrogen just above 300 nm removes both carbonyl and hydroperoxy groups, but exposure above 370 nm removes only the latter. Photooxidation of control and pretreated samples was then carried out in air using a 500 W mercury/tungsten lamp. Results of these experiments are shown in Figure B-5. As anticipated, carbonyl and hydroperoxy groups in the control sample were found to strongly influence its photooxidation rate. Destruction of these active groups prior to exposure, however, does not show any significant change in stability. This result has led the authors to question the role of both hydroperoxy and carbonyl groups as primary photochemical initiators. As an alternate mechanism, they have suggested that oxygen/polymer, charge-transfer complexes (III) may be important in the initiation step,

$$RH + O_2 \rightarrow [R{-}H{-}{-}{-}O_2] \xrightarrow{hv} [RH^+ {-}{-}{-}O_2^-] \rightarrow R^\bullet + HOO^\bullet .$$

(III)

In a hydrocarbon polymer, free of carbonyl and hydroperoxy groups, such complexes could be the primary initiators. However, once photooxidation has been initiated and active groups begin to accumulate, hydroperoxides probably account for most of the reaction, at least in the early stages of degradation. As photooxidation reaches the more advanced stage, carbonyl groups probably become the predominant uv absorbing species.

3) Photooxidation of Blends of Polyethylene and Polypropylene

Interesting effects are observed when blends of two hydrocarbon polymers are oxidized. For example, completely compatible blends of 5% polypropylene (PP) with LDPE photooxidize more rapidly than polyethylene alone [72]. Similarly, addition of an ethylene/propylene copolymer as a compatibilizing agent to a 1:1 blend of PP and LDPE significantly increases the rate of photooxidation. Scott and coworkers [72-76] have attributed this loss of stability to prooxidant effects of chain branching and unsaturation in the photooxidation of these blends. Apparently there is a chemical interaction between the two phases during photooxidation. There is also evidence supporting interfacial grafting at the LDPE/PP boundaries. Reactive species, generated in the less stable component, should be expected to induce degradation reactions in the more stable polymer.

4) Thermal and Photooxidation of Polystyrene

Geuskens [77] has studied the effect of processing on the photooxidative stability of polystyrene (PS). Thermal oxidation in a Brabender Plastograph at 160 °C is indicated by a decrease in tensile strength with increasing time of processing. Groups capable of absorbing uv radiation are formed when air is present, and the polymer is then sensitized to photooxidation. Hydroperoxides (IV) are believed to be the first molecular products formed, and these then decompose to form ketone groups as follows,

$$---CH_2-\underset{\underset{IV}{|}}{\overset{\overset{OOH}{|}}{C}}-CH_2-CH--- \quad \xrightarrow{uv} \quad ---CH_2-\overset{\overset{O}{\|}}{C} \; + \; H_2O \; + \; CH=CH---$$

Thermal oxidation is negligible when PS is processed in the absence of air, and the limited decrease in molecular weight which is observed can be attributed to shear forces. Groups that absorb uv radiation are not produced in the absence of air so processing under these conditions does not sensitize PS to photooxidation. However, the complete exclusion of air in commercial processing is seldom achieved.

Phenolic antioxidants[1] are effective inhibitors for the thermal oxidation of PS, but hindered amines[1] are much less effective. Yet processing of PS containing either type

---

1 These antioxidants are described in Section D

of antioxidant stabilizes the polymer against subsequent photooxidation. The phenolic antioxidants inhibit thermal oxidation whereas hindered amines function as photostabilizers, reducing the quantum yield of photooxidation [78]. There is little change in tensile strength of PS processed with phenolic antioxidants added indicating negligible thermal oxidation, but degradation still takes place during subsequent exposure to uv. In contrast, tensile strength of PS protected with an amine antioxidant is reduced during processing. Hindered amines function as stabilizers in the subsequent photooxidation. Combinations of amine and phenolic antioxidants give maximum protection.

Ränby [79] has confirmed the presence of hydroperoxides in photooxidized PS by infrared spectroscopy. Alkyl and peroxy radicals are the first products which form after abstraction of a hydrogen from a tertiary carbon atom. Homolysis of hydroperoxides yields alkoxy radicals (V) which may then decompose by the following reactions,

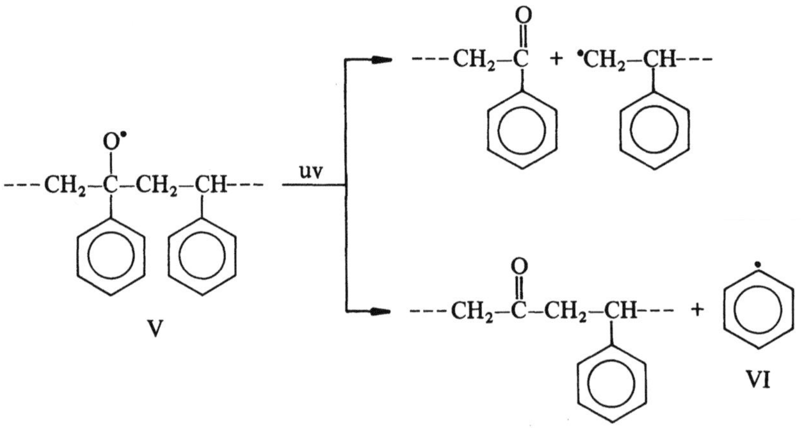

Support for the decomposition into phenyl radicals (VI) and aliphatic ketone groups was obtained from parallel reactions with the model compound, 2-phenylbutane in which phenol was identified among the reaction products. Although in the model compound infrared spectroscopy of the reaction products showed an absorption band characteristic of phenol, this band was apparently hidden in photooxidized PS by the accompanying polymer alkyl and polymer peroxy radicals.

Despite a period of uncertainty, it is now generally agreed that hydroperoxides play an important role in the photooxidation of polymers [80, 81]. The recent studies on mechanisms for the photooxidation of PS and the role of hydroperoxides in this degradation process have made an important contribution toward establishing general mechanisms for photooxidation.

5) Degradation of Bisphenol-A-Polycarbonate

Polycarbonates undergo thermal degradation by a radical-initiated, random-chain-scission mechanism [82]. It has also been suggested that photooxidation of these polymers may involve similar free-radical reactions. However, the sensitivity of polycarbonates to uv radiation suggests the possibility of a photo-Fries rearrangement as a component of the overall mechanism [83-85]. This rearrangement has been observed with the model compound, bis(4-tert.-butylphenyl) carbonate (VII).

VII

Where ─┼─ is tert.butyl

Loss in mechanical strength as well as discoloration could be accounted for by such a mechanism. Similar reactions may take place on exposure to uv radiation in a nitrogen atmosphere. Though evidence has been presented indicating that this rearrangement occurs in solution, it has not been confirmed during photooxidation in the solid state.

An alternative mechanism for photodegradation of polycarbonates is based on oxidation reactions [86-88]. Recently, Factor and Chu [89] have confirmed the important role which oxidation plays in the photodegradation of polycarbonates during outdoor weathering, particularly under accelerated exposure conditions. Reaction with oxygen is slow at first but rapidly becomes autocatalytic. These investigators have proposed a complex mechanism based on free-radical reactions. In this mechanism, both photo-rearrangement and oxidation reactions are included. Many of the products and intermediates suggested have been identified, and these are underlined in the proposed mechanism.

Hydrolytic reactions also contribute to the photodegradation of polycarbonates based on Bisphenol A [90,91]. These polymers are reasonably stable in dry air [92]. Discoloration does occur eventually, but mechanical properties undergo very little change, unless moisture is present.

Possible Mechanism of PC Photooxidation [89]

Primary Processes
  Photo-Fries:

$$PC \xrightarrow{uv} ArO^{\cdot} + \quad \xrightarrow{O_2} RO_2H + \underline{quinone} + \underline{salicylate}$$

O_2–Charge Transfer Absorption:

$$RCH_3 + O_2 \underset{uv}{\rightleftharpoons} [RH^{\oplus}{\cdots}O_2^{\ominus}]^* \longrightarrow [R{\cdots}O_2H] \longrightarrow RO_2H$$

B) Polymer Degradation

Metal Impurities:

$$Fe^{3+}OH^- \xrightarrow{uv} Fe^{2+} + \cdot OH$$

Secondary Processes

$$RO_2H \xrightarrow{uv} RO\cdot + \cdot OH$$

$$RCOR \xrightarrow{uv} \begin{cases} RCO\cdot + R\cdot \xrightarrow{O_2} RO_2H \\ [RCOR]^* \xrightarrow{RH} R\overset{OH}{\underset{\cdot}{C}}R + R\cdot \end{cases}$$

$$Ar_2CMe_2 \xrightarrow{R\cdot} Ar_2\overset{CH_2\cdot}{\underset{|}{C}}{-}Me \xrightarrow{O_2} Ar_2\overset{CH_2O_2H}{\underset{|}{C}}Me \longrightarrow Ar_2\overset{CH_2OH}{\underset{|}{C}}Me$$

$$\downarrow O \qquad\qquad \downarrow H^+ \qquad\qquad \downarrow$$

$$Ar\overset{CH_2}{\underset{|}{C}}Me \qquad \boxed{\textit{acids, esters, anhydrides}}$$

$$\downarrow O_2$$

$$Ar\overset{CH_2}{\underset{|}{C}}{-}Me \qquad Ar\overset{O}{\overset{\|}{C}}Me + ArCH_2OH \longrightarrow ArCO_2H$$

$$\overset{|}{O_2H} \longrightarrow ArOH + ArCH_2\overset{O}{\overset{\|}{C}}Me \longrightarrow ArCH_2\overset{O}{\overset{\|}{C}}OH$$

Products and intermediates identified in this or other work are underlined

6) Thermal and Photooxidation of Polyacetals

Unless stabilized with carbon black, poly(phenylene oxide) (PPO) has poor stability to outdoor weathering. This polymer degrades by photooxidation at wavelengths as high as 365 nm [93], and its degradation is affected by both thermal oxidation and moisture [94]. Hydroperoxides, formed during processing, sensitize PPO to photooxidation [95]. Davis [94] has made an extensive study of meteorological factors on the degradation of PPO at several geographic locations. Degradation was monitored by weight loss occurring over a six month, summer interval. The following relationship was developed from these data,

$$W = 1.58 \times 10^4 De^{-10,000RT}$$

in which W is the weight loss, D the annual uv dose in BLETs, e the activation energy, and T the mean absolute temperature. This relationship was then used to predict the weight loss of PPO during outdoor weathering at locations throughout the world using the appropriate meteorological data. Data obtained at fourteen test sites confirmed this relationship.

It has been reported [96, 97] that moisture inhibits the photooxidation of PPO and polyacetals. Presumably this occurs by leaching out of prodegradants including formic and acetic acid which are products of the degradation.

The infrared spectrum of commercial polyoxymethylene (POM) shows the presence of several types of carbonyl impurities [98, 99]. Unsaturated aldehydic carbonyls and α,β-unsaturated carbonyl groups in particular, are responsible for initiating the photooxidation of POM. Saturated aldehydic groups also are formed, but only in the final stages of photooxidation [100–102]. It is suggested [99] that bond scission in POM occurs through the following mechanism involving terminal aldehydic groups (VIII).

$$- -CH_2-O-CH_2-O-CHO \xrightarrow{h\nu} - - -CH_2-O-CH_2-O^{\bullet} + {}^{\bullet}CHO$$

$$\text{VIII} \qquad\qquad\qquad\qquad \text{IX}$$

Formaldehyde would then form by hydrogen abstraction by the ${}^{\bullet}CHO$ radical,

$$- -CH_2-O-CH_2-O- - - - + {}^{\bullet}CHO$$

$$\text{IX}$$

$$\rightarrow - -CH_2-O-\overset{\bullet}{C}HO- - - - + H_2C{=}O$$

The polymer radical (IX) could then undergo further reaction as follows,

$$- -CH_2-O-CH_2-O^{\bullet} \begin{cases} \xrightarrow{+ POM} - -CH_2-O-CH_2-OH + POM^{\bullet} \\ \\ \xrightarrow{} - -CH_2-O-CH + H^{\bullet} \\ \qquad\qquad\qquad\qquad \underset{O}{\overset{\|}{}} \end{cases}$$

The α,β-unsaturated carbonyl groups may also contribute to chain scission by the following reactions,

$$\underset{\overset{\|}{O}}{- -O-\overset{H}{C}{=}\overset{H}{C}-C-CH_2-O- - -} \xrightarrow{h\nu} \underset{\overset{\|}{O}}{- -O-\overset{H}{C}{=}\overset{H}{C}-C^{\bullet} + {}^{\bullet}CH_2-O- -}$$

$$\underset{\overset{\|}{O}}{- -O-\overset{H}{C}{=}\overset{H}{C}-C-O-CH_2- - -} \xrightarrow{h\nu} \underset{\overset{\|}{O}}{- -O-\overset{H}{C}{=}\overset{H}{C}-C^{\bullet} + {}^{\bullet}O-CH_2- -}$$

IV) Degradation by Hydrolysis

Those polymers that are synthesized by condensation reactions are particularly susceptible to degradation by hydrolysis. Random scission of bonds along the backbone chain can occur as in,

$$R-N-\underset{\overset{\|}{O}}{\overset{H}{|}}C-R^{\bullet} \xrightarrow{H_2O} R-NH_2 + HO-\underset{\overset{\|}{O}}{C}-R^{\bullet}$$

Poly(methyl methacrylate), on the other hand, is an addition-type polymer which undergoes hydrolytic degradation without chain scission, reaction taking place at ester bonds on substituent groups. Hydrolysis may be catalyzed by either acids or bases, and additional catalysts may form within the polymer as a result of accompanying thermal oxidation.

Polyamides, polyesters, polycarbonates, and polysaccarides are among the important, commercial polymers that are degraded by hydrolysis. As in other types of degradation discussed previously, there are important differences among polymers in their stability to hydrolysis. Nylon 11 has a longer hydrocarbon segment than 6,6 nylon and is therefore more resistant to hydrolytic degradation. The rate of hydrolysis is also limited by the diffusion of water into the polymer bulk. For significant reaction to occur, water must be absorbed at the surface and then permeate into subsurface regions. Flexibility of polymer chains also contributes to hydrolytic instability [103] by opening the structure to moisture penetration. Hydrolysis is believed to occur primarily in amorphous regions [104] which are more readily penetrated by water molecules.

References

1. Howard, J.B.: Stress-cracking, in: Crystalline Olefin Polymers (eds) Raff, R.A.V., Doak, K.W., pp. 47, New York: Interscience 1964
2. Heiss, J.H., Lanza, V.I.: Wire Prod., *33*, 1182 (1958)
3. Williams, C.G.: Proc. Roy. Soc. (London), *10*, 516 (1860)
4. Madorsky, S.L.: Thermal Degradation of Polymers, New York: Interscience 1964
5. Grassie, N.: Chemistry of High Polymer Degradation Processes, London: Butterworth 1956
6. Grassie, N.: Encyclopedia of Polymer Science and Technology, New York: Interscience 1966
7. Chiatore, O. et al.: Poly. Degrad. and Stab., *3 No. 3*, 209 (1981)
8. Nurayama, N., Amagi, Y.: J. Polym. Sci., Part B, *4*, 115 (1966)
9. Mitani, K. et al.: J. Polym. Sci., Polym. Chem. Ed., *13*, 2813 (1975)
10. Suzuki, Y., Takakura, I., Yoda, M.: Eur. Polym. J., *7*, 1105 (1971)
11. Mayer, Z.: J. Macromol. Sci., *10*, 263 (1974)
12. Valko, L., Tvaroski, P.: Eur. Polym. J., *11*, 411 (1975)
13. Starnes, W.H., Jr.: Recent Fundamental Developments in the Chemistry of (PVC) Degradation and Stabilization, in: Stabilization and Degradation of Polymers (eds.) Allara, D.L., (ed) Hawkins, W.L., p. 309, Advances in Chemistry Series, Amer. Chem. Soc. 1978
14. Braun, D., in: Degradation and Stabilization of Polymers (ed) Geuskens, G., p. 23, New York: Wiley 1975
15. Plitz, I.M., Willingham, R.A., Starnes, W.H., Jr.: Macromol., *10*, 499 (1977)
16. Papko, R.A., Pudov, V.S.: Polym. Sci. (USSR), *16*, 1636 (1974) (English Translation)
17. Gupta, V.P., St. Pierre, L.E.: J. Polym. Sci., Polym. Chem. Ed., *11*, 1841 (1973)
18. Dodson, B., McNeill, I.C.: J. Polym. Sci., Polym. Chem. Ed., *14*, 353 (1976)
19. Hoang, T.V. et al.: Eur. Polym. J., *11*, 469 (1975)
20. Zafar, M.M., Mahmood, R.: Eur. Polym. J., *12*, 333 (1976)
21. Varma, I.K., Grover, S.S.: Macromol. Chem., *175*, 2515 (1974)
22. Abdullin, M.I. et al.: Preprints from the Second Internat. Symp. on PVC, Lyon-Villeurbonne, France, p. 272 (1976)
23. Ayrey, G., Head, B.C., Poller, R.C.: J. Polym. Sci. Macromol. Rev., *8*, 1 (1974)
24. David, C.: Compr. Chem. Kinet., *14*, 78 (1975)
25. MacDonald, R.N.: U.S. Patent 2,768,994 (1956)
26. Allen, N.S., McKellar, J.F.: Polym. Degrad. and Stab., *1 No. 1*, 47 (1979)
27. Cameron, G.C., Keer, G.P.: Europ. Polym. J., *4*, 709 (1968)

28. Loan, L.D., Winslow, F.H., Reactions of Macromolecules, in: Macromolecules (eds) Bovey, F.A., Winslow, F.H., p. 433, New York: Academic Press 1979
29. Madorsky, S.L., Straus, S.J.: J. Res. Natl. Bur. Stand, *53*, 361 (1954)
30. Wall, L.A., Straus, S.J.: J. Polym. Sci., *43*, 313 (1960)
31. Chien, J.C.W., Kiang, J.K.Y., Pyrolysis and Oxidative Pyrolysis of Polypropylene, in: Stabilization and Degradation of Polymers (eds) Allara, D.L., Hawkins, W.L., p. 175, Advances in Chem. Series, Amer. Chem. Soc. 1978
32. Kiang, J.K.Y., Uden, P.C., Chien, J.C.W.: Polym. Degrad. and Stab., *2 No. 2*, 113 (1980)
33. Grassie, N.: Soc. Chem. Ind. Monograph, *26*, 191 (1967)
34. McNeill, I.C., The Thermal Degradation of Polymer Blends, in: Developments in Polymer Degradation (ed) Grassie, N., p. 171, London: Applied Science Publishers 1977
35. McNeill, I.C., Gupts, S.N.: Polym. Degrad. and Stab., *2 No. 2*, 95 (1980)
36. Grassie, N., Davidson, A.J.: Polym. Degrad. and Stab., *3 No. 3*, 25 (1980)
37. Hawkins, W.L., Environmental Deterioration of Polymers, in: Polymer Stabilization (ed) Hawkins, W.L., p. 1, New York: Wiley-Interscience 1972
38. Winslow, F.H. et al.: Chem. Ind., 533 (1963)
39. Shelton, J.R., Stabilization Against Thermal Oxidation, in: Polymer Stabilization (ed) Hawkins, W.L., p. 1, New York: Wiley-Interscience 1972
40. Bateman, L.: Quart. Rev. (London), *8*, 147 (1954)
41. Barnard, D. et al., Oxidation of Olefins and Sulfides, in: Chemistry and Physics of Rubber-like Substances (ed) Bateman, L., p. 593, London: Macheren 1963
42. Shelton, J.R., Stabilization Against Thermal Oxidation, in: Polymer Stabilization (ed) Hawkins, W.L., p. 29, New York: Wiley-Interscience 1972
43. Chan, M.G., Hawkins, W.L.: Polym. Eng. Sci., 3 (1967)
44. Hansen, R.H., Martin, W.H., DeBenedictis, T.: Trans. Inst. Rubber Ind., *39*, p. 301 (1963)
45. Wall, L.A., Harvey, M.R., Tryon, M.J.: J. Phys. Chem., *60*, 1306 (1956)
46. Keith, H.D., Padden, F.J., Jr.: J. Appl. Phys., *42*, 4585 (1971)
47. Hirt, R.C., Searle, N.Z., Schmitt, R.G.: Soc. Plastics Engs., Trans., *1*, 21 (1961)
48. Pross, A.W., Black, R.M.: J. Chem. Ind. (London), *69*, 113 (1950)
49. Burgess, A.R.: Natl. Bur. Stds. (USA), Circular, 525, 149 (1953)
50. Hartley, G.H., Huillet, J.: Macromol., *1*, 165, 413 (1968)
51. Trozzola, A.M.: Stabilization Against Oxidative Photo-Degradation, in: Polymer Stabilization (ed) Hawkins, W.L., p. 159, New York: Wiley-Interscience 1972
52. Carlsson, D.J., Wiles, D.M.J.: Macromol. Sci., *C14*, 65 (1976)
53. Guillet, J.: Fundamental Processes in the Photodegradation of Polymers, in: Stabilization and Degradation of Polymers (eds) Allara, D.L., Hawkins, W.L., p. 1, Advances in Chemistry Series, Amer. Chem. Soc. 1978
54. Black, R.M.: J. Appl. Chem., *8*, 159 (1958)
55. Black, R.M., Lyons, B.J.: Proc. Roy. Soc. (London), *253A*, 322 (1959)
56. Bovey, F.A., Schilling, F.C., Cheng, H.N.: ^{13}C NMR Observation of the Effects of High Energy Radiation and Oxidation on Polyethylene and Model Paraffins, in: Stabilization and Degradation of Polymers (eds) Allara, D.L., Hawkins, W.L., p. 134, Advances in Chemistry Series, Amer. Chem. Soc. 1978
57. Criegee, R.: Rec. Chem. Progr., (Kresge-Hocker Lib.), *18*, 111 (1957)
58. Staudinger, H.: Ber., *58*, 1088 (1925)
59. Bailey, P.S., Thompson, J.A., Shoulders, B.A.: J. Amer. Chem. Soc., *88*, 4098 (1966)
60. Murray, R.W.: Prevention of Degradation by Ozone, in: Polymer Stabilization, p. 215, (ed) Hawkins, W.L., New York: Wiley-Interscience 1972
61. Kaufman, F.S.: A New Technique for Evaluating Outdoor Weathering Properties of High Density Polyethylene, in: Applied Polymer Symposia, *No. 4*, 131 (1967) New York: Wiley-Interscience
62. Carlsson, D.J., Garton, A., Wiles, D.M.: Some Effects of Production Conditions on the Photosensitivity of Polypropylene Fibers, in: Stabilization and Degradation of Polymers (eds) Allara, D.L., Hawkins, W.L., p. 56, Advances in Chemistry Series, Amer. Chem. Soc. 1978
63. Howard, J.B.: Stress Cracking, in: Crystalline Olefin Polymers (eds) Raff, R.A.V., Doak, K.W., p. 47, New York: Wiley-Interscience 1964

B) Polymer Degradation

64. Howard, J.B., Gilroy, H.M.: Soc. Plastics Engs. J., *24(1)*, 68 (1968)
65. Kambour, R.P.: Polym., *5*, 143 (1959)
66. Nielsen, L.E.: J. Appl. Polym. Chem., *1*, 24 (1959)
67. Carlsson, D.J., Wiles, D.M.: J. Polym. Sci., Polym. Letters (ed) *8*, 419 (1970)
68. Furneaux, G.C., Ledbury, K.J.: Polym. Degrad. and Stab., *3 No. 6*, 431 (1981)
69. Davis, A., Deane, G.H.W., Duffy, B.L.: Nature, *18*, 1159 (1976)
70. Allen, N.S.: Polym. Degrad. and Stab., *2 No. 2*, 155 (1981)
71. Allen, N.S., Fatinikun, K.O.: Polym. Degrad. and Stab., *3 No. 5*, 327 (1981)
72. Sadrmohaghegh, C., Scott, G., Setoudeh, E.: Polym. Degrad. and Stab., *3 No. 6*, 469 (1981)
73. Chakraborty, K.B., Scott, G.: Polym. Degrad. and Stab. *1*, 37 (1979)
74. Ghaffar, K.B., Scott, A., Scott, G.: Eur. Polym. J., *12*, 615 (1976)
75. Scott, G., Tahan, M.: Eur. Polym. J., *13*, 989 (1977)
76. Scott, G.: Adv. in Chem. Series, *169*, 30 (1978)
77. Geuskens, G. et al.: Polym. Degrad. and Stab., *3 No. 4*, 295 (1980–81)
78. Carlsson D.J. et al.: Pure Appl. Chem., *52*, 389 (1980)
79. Lucki, J., Ranby, B.: Polym. Degrad. and Stab., *1 No. 1*, 1 (1979)
80. Bolland, J.L., Gee, G.: Trans. Faraday Soc., *42*, 244 (1946)
81. Rabek, J.F., in: Comprehensive Chemical Kinetics (eds) Bamford, C.H., Tipper, V.F., Vol. *14*, 425, Elsevier 1974
82. Davis, A., Golden, J.H.: Macromol. Chem., *78*, 16 (1964)
83. Bellus, D., Hardlovic, P., Manasek, Z.: Polym. Letters, *4*, 1 (1966)
84. Humphrey, J.S., Jr., Shultz, A.R., Jacquiss, D.G.B.: Macromol., *6 No. 3*, 305 (1973)
85. Mullen, P.A., Searle, N.Z.: J. Appl. Polym. Sci.: *14*, 765 (1970)
86. Sato, Y.: Polym. Chem. (Japan), *21*, 232 (1964)
87. Gesner, B.D., Kelleher, P.D.: J. Appl. Polym. Sci.: *13*, 2183 (1969)
88. Humphrey, J.S., Jr., Roller, R.S.: Mol. Photochem., *3*, 35 (1971)
89. Factor, A., Chu, M.L.: Polym. Degrad. and Stab., *2*, 203 (1980)
90. Davis, A., Golden, J.H.: J. Macromol. Sci., *C3*, 49 (1969)
91. Gardner, R.J., Martin, J.R.: J. Appl. Polym. Sci., *24*, 1269 (1979)
92. Abbas, K.B.: Preprints International Conference on Plastics in Telecommunications II, Plastics and Rubber Institute (London) 1978
93. Grassie, N., Roche, R.S.: Die Macromol. Chem., *112*, 34 (1968)
94. Davis, A.: Polym. Degrad. and Stab., *3*, 187 (1980)
95. Chakraborty, K.B., Scott, G.: Eur. Polym. J., *15*, 35 (1979)
96. Faidel, G.I., Vol'fson, S.A.: Sov. Plastics, *8*, 29 (1967)
97. Ivanova, L.V. et al.: J. Polym. Sci. USSR, *A16 No. 8*, 1831 (1974)
98. Fox, R.B., Isaacs, L.G., Stokes, S.: J. Polym. Sci., Chem. (ed) *1*, 1079 (1963)
99. Allen, N.S., McKellar, J.F.: Polym. Degrad. and Stab., *1 No. 1*, 47 (1979)
100. Grassie, N., Roche, R.S.: Macromol. Chemie, *112*, 34 (1964)
101. Kelleher, P.G., Jassie, L.B.: J. Appl. Polym. Sci., *9*, 2501 (1965)
102. Hughes, O.R., Coard, L.C.: J. Polym. Sci., Chem. (ed) *7*, 1861 (1969)
103. Goldberg, E.P.: J. Polym. Sci., Part C, Polym. Symp. *4*, 707 (1963)
104. Ravens, D.A.S., Sisley, J.E.: Hydrolysis, in: Chemical Reactions of Polymers (ed) Fettes, E.M., p. 551, New York: Wiley-Interscience 1964

C) Stabilization Against Non-oxidative Thermal Degradation

There are two general approaches to the stabilization of polymers; (1) by modification of molecular structure and (2) by the use of additives. Dependent on the mode of degradation, either or both of these techniques may be employed. When the initiation rate is low, additives can be effective and radical traps[1] or chain terminators[1] may be used under these conditions. At high initiation rates, additives are less effective, and stabilization by structure modification may be the only practical approach. Additives would be overwhelmed by the large number of radicals formed during initiation.

Thermal degradation in the absence of oxygen (pyrolysis) usually takes place at elevated temperatures. The initiation rate is high and therefore additives are not nearly so effective as in the inhibition of low-temperature oxidation or photooxidation. Structural modification in which relatively weak bonds are replaced by bonds of higher dissociation energy is the preferred approach. As an example, polytetrafluoroethylene with only C—C and C—F bonds in the structure is much more stable to pyrolysis than is polyethylene. Replacement of C—H bonds with C—F bonds, however, results in a polymer quite different from polyethylene in many other important aspects. The extent to which a specific polymer can be heat stabilized by structural modification or by copolymerization is limited because other properties of the original polymer must usually be held within closely specified limits.

I) Stabilization by Structural Modification

There is wide variation in the stability of polymers to nonoxidative thermal degradation[1]. This can be attributed to differences in the dissociation energies of those bonds which make up the structure. At elevated temperatures, bonds may cleave along the backbone chain in a random-scission process or by depolymerization. In other polymers, substituents are split off from the backbone chain. Whatever the mechanism of pyrolysis, it is evident that replacement of weak bonds with others having a higher dissociation energy will increase thermal stability. This approach has been applied successfully in a limited number of cases in which weak links or labile substituents are responsible for initiating degradation. Replacement of a few such bonds can usually be accomplished without altering the original polymer beyond acceptable limits.

It has long been suspected that structural defects in poly(vinyl chloride) (PVC) are responsible for the poor thermal stability of this important polymer. Various investigators have suggested terminal chlorine, alkene groups, branch points, head-to-head

---

1 These stabilizers are described in Sect. D

structures, and irregularities formed during prior oxidation as potential initiation sites. Identification of the exact defect structures responsible for initiation of the thermal degradation of PVC, however, has not as yet been established. Several recent reviews [2-5] deal with the relative importance of suggested structural defects on the thermal degradation of PVC, and the reader is directed to these for a more complete discussion of this controversial subject. Though the weight of evidence supports structural defects as the primary initiation sites, the possible contribution of normal repeating units cannot be neglected [5-8]. Nonetheless, prior reactions of PVC which could remove labile groups is an attractive possibility for stabilization in that the splitting off of hydrogen chloride would be suppressed. This is in contrast to the more conventional mechanisms for stabilization of PVC in which basic additives are used to neutralize hydrogen chloride which is now generally agreed to be a catalyst for PVC degradation.

In 1959 Frye and Horst [9] proposed a mechanism for stabilization of PVC using organometallic stabilizers. However, attempts to demonstrate the addition of ligands from the stabilizer to PVC molecules were not successful at first, and the theory did not find wide acceptance. Starnes and coworkers [10, 11] reasoned that insufficient stabilizer had been used in previous experiments and that the reaction might not proceed readily in the molten polymer. These investigators studied the reaction of PVC in o-chlorobenzene solution with a large excess of di(n-butyl)bis(n-dodecylthio)stannate as the stabilizer. Sufficient stabilizer was used to react with all of the reactive sites believed to be present, and reactions were carried out at temperatures up to 185 °C. Under these conditions, it was found that reactions proposed by Frye and Horst did indeed take place as follows,

$$
\text{-----} \overset{}{C}\!=\!\overset{}{C}\!-\!\overset{\overset{\textstyle Cl}{|}}{CH}\!-\!CH_2 \text{-----} + MY_2 \rightarrow
$$

$$
\rightarrow \text{-----} \overset{}{C}\!=\!\overset{}{C}\!-\!\overset{\overset{\textstyle Y}{|}}{CH}\!-\!CH_2 \text{-----} + MYCl
$$

in which M is $(C_2H_5)_2Sn^{+2}$ and Y is $C_{12}H_{25}S^-$. Similar results were obtained with Ba^{+2}, Cd^{+2}, Zn^{+2}, Pb^{+2} as the metal and with RS^-, RCO_2 or RO^- as the organic anion. Thermal stability of the modified PVC was increased up to ninefold by replacement of labile chlorines, and stability increased with the amount of ligand bonded to the polymer. Thus structural modification of PVC has been shown to be an effective method for stabilizing this sensitive polymer against thermal degradation. Although recent evidence has been reported suggesting the presence of weak links in other polymers, as in radical-polymerized polystyrene [12], these bonds appear to be in the backbone chain. Replacement of such weak links would require a major alteration in structure of the polymer.

II) Stabilization by Copolymerization

There are several important examples in which thermal stability has been improved by copolymerization of the primary monomer with traces of a comonomer. Endcapping of the terminal hydroxyl group in polyoxymethylene (POM) is one such example having considerable commercial importance.

Formaldehyde reacts spontaneously to form a low-molecular-weight oligomer. Although it had been known for some time that macromolecules could be produced from formaldehyde, the homopolymer degraded rapidly once removed from the reaction environment. Depolymerization is initiated at chain ends as indicated by the inverse relationship [13] between stability and molecular weight in these polymers. As a result, the excellent mechanical properties anticipated in the high-molecular-weight polymers could not be realized. In 1956, however, reactions were developed [14] in which the terminal hydroxyl group was converted to either an ester or an ether group. Depolymerization was inhibited by this structural modification, and polyoxymethylenes became available commercially [15, 16].

Reaction with acetic acid anhydride is typical of this structural modification,

$$----CH_2-O-CH_2-OH \ + \ (CH_3CO)_2) \ \rightarrow \ ----CH_2-O-CH_2-OCOCH_3 + CH_3COOH.$$

Although these reactions prevented depolymerization initiated at chain ends, random scission could still take place. Depolymerization would then proceed from the point of random cleavage with eventual reversion back to monomer. To minimize this problem, formaldehyde has been copolymerized [17] with small amounts of various comonomers including ethylene oxide. Even when random chain cleavage occurs in these copolymers, depolymerization stops when the first comonomer unit is reached. Although there would be some degradation with a loss in mechanical strength, the effect would not be as severe as in POM polymers with only endcapping. Modified polyoxymethylenes are used in many applications because of their superior strength and clarity.

Poly(methyl methacrylate) (PMMA) is an important polymer used extensively for decorative applications and as a glazing material. The homopolymer, however, degrades rapidly at elevated temperatures, depolymerizing to monomer in 95% yield. Poly(methyl acrylate) (PMA), on the other hand, is much more resistant to depolymerization. Random copolymers of methyl methacrylate and methyl acrylate should have better thermal stability than PMMA since depolymerization would not proceed as readily once a methyl acrylate unit is reached. Copolymers with varying amounts of the two monomers have been made [18], and although the anticipated improvement in stability was realized, other important properties of PMMA were sacrificed.

III) Stabilization by Crosslinking

The bonds forming crosslinks in network polymers contribute significantly to thermal stability. Thus thermosets as a general class have greater resistance to heat than do the thermoplastics. Though this approach to stabilization is one, more of polymer design than modification, crosslinking can be considered as a special type of structural modification.

The excellent heat resistance of melamine-formaldehyde polymers enables these thermoset resins to be used in applications such as counter tops where high temperatures may be encountered. Thermosets formed from formaldehyde and urea or phenol also have excellent thermal stability. Surface crosslinking of poly(vinyl chloride) by

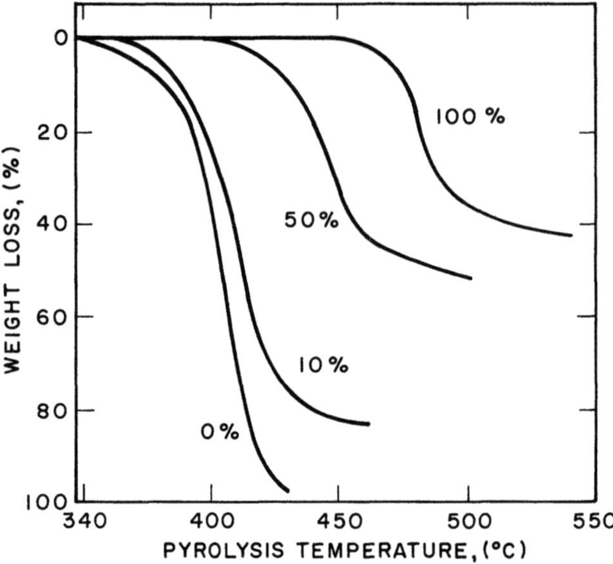

Fig. C-1. Pyrolysis patterns of styrene-trivinylbenzene copolymers designated by the trivinylbenzene concentration; rate of temperature rise, 100 °C/h. (Reprinted with permission of Wiley-Interscience)

radiation forms a product having both outstanding mechanical strength and good thermal stability.

Some polymers designed for high-temperature resistance take advantage of the inherent stability of network structures. When a single covalent bond in a thermoplastic is broken, chain scission occurs, but in the densely crosslinked structures many bonds must be broken before extensive degradation takes place. Thermal stability increases with crosslink density as shown in Fig. C-1. Here the stability of a series of styrene-trivinylbenzene copolymers is plotted against the extent of crosslinking which takes place through the trivinylbenzene units.

Polymers with very high heat resistance have been used as ablation shields to protect space vehicles during re-entry into the earth's atmosphere. Many of these specialized polymers have a highly cross-linked or ladder structure [19]. A typical structure for a ladder polymer based on silicon would be,

Heat resistant polymers are designed to withstand temperatures above 300 °C for extended periods and much higher temperatures for shorter exposures.

38

IV) Stabilization with Additives

As previously noted, thermal degradation in the absence of oxygen usually takes place at higher temperatures than oxidative degradation. The initiation rate is higher at elevated temperatures and as a result additives are less effective. Added stabilizers would require high mobility to reach sites of incipient degradation, and would have to be present in high concentration in order to react with the large number of radicals produced under these conditions. Despite these restrictions, additives can perform at least a minor role in stabilization against thermal degradation.

Additives may inhibit thermal degradation by modifying relatively weak bonds under elevated temperature conditions to form bonds with higher dissociation energies. Such reactions are highly specialized and have only limited application. These reactions would be different from prior treatment of poly(vinyl chloride) to replace labile chlorines. Alternatively, an additive could function as a chain terminator. In thermal degradation, however, chain reactions do not play the dominant role that they do in low-temperature oxidations. Additives are more useful when oxidation takes place simultaneously with thermal degradation.

References

1. Loan, L.D., Winslow, F.H., in: Polymer Stabilization (ed) Hawkins, W.L., p. 140, New York: Wiley-Interscience 1972
2. Ayrey, G., Head, B.C., Poller, R.C.: J. Polym. Sci., Macromol. Rev., *8*, 1 (1974)
3. Mayer, Z.: J. Macromol. Chem., *10*, 263 (1924)
4. David, C.: Compr. Chem. Kinet., *14*, 78 (1975)
5. Nass, L.O.: Encycl. PVC, *1*, 271 (1976)
6. Abbos, K.B., Sorvic, E.M.: J. Appl. Polym. Sci., *20*, 2395 (1976)
7. Troitski, B.B. et al.: Eur. Polym. J., *11*, 277 (1975)
8. Nolan, K.P., Shapiro, J.S.: J. Chem. Soc., Chem. Commun., 490 (1075)
9. Frye, A.H., Horst, R.W.: J. Polym. Sci., *40*, 419 (1959)
10. Plitz, I.M., Willingham, R.A., Starnes, W.H., Jr.: Macromol., *10*, 499 (1977)
11. Starnes, W.H., Jr., Recent Fundamental Developments in the Chemistry of Poly(vinyl chloride) Degradation, in: Stabilization and Degradation of Polymers, (eds) Allara, D.L., Hawkins, W.L., ACS Advances in Chemistry Series *169*, 309 (1978)
12. Chiantore, O. et al.: Polym. Degrad. and Stab., *3 No. 3*, 209 (1981)
13. Grassie, N., Roche, R.S.: Makromol. Chem., *112*, 16 (1968)
14. MacDonald, R.N.: US Patent 2,768,994 (1956)
15. Schweitzer, C.E., MacDonald, R.N., Punderson, J.O.: J. Appl. Polym. Sci., *1*, 158 (1959)
16. Koch, T.A., Lindvig, P.E.: J. Appl. Polym. Sci., *1*, 164 (1959)
17. Walling, C.T., Brown, F., Bartz, K.W.: US Patent 3,027,352
18. Grassie, N., Torrance, B.J.D.: J. Polym. Sci., Part A-1, *6*, 3303 (1968)
19. DeWinter, W.: J. Macromol. Sci., Rev. Macromol. Chem., *1*, 329 (1966)

D) Stabilization Against Thermal Oxidation

The autoxidation of hydrocarbons is probably the most extensively studied of all chemical reactions. This statement has been attributed to Bateman [1] whose pioneering research at the British Rubber Producer's Association between 1954 and 1964 established the mechanism for the low-temperature oxidation (autoxidation) of low-molecular weight hydrocarbons as models for natural rubber. Accepting the validity of this conclusion, today it may be said that reactions responsible for stabilization of hydrocarbons and hydrocarbon polymers have been even more extensively studied.

The search for stabilizers to protect polymers against autoxidation had been underway many years before Bateman and his associates [2-3] proposed their mechanism for the autoxidation of rubber. As early as 1937, Semon [4] showed that phenolic compounds retarded the autoxidation of rubber, and Ostwald [5] reported inhibition of this degradation by aromatic amines. There then followed a series of papers by Moureu and Dufraise [6,7] who developed one of the earliest theories for antioxidant action. This theory is often referred to as the "negative catalyst" theory [8] in which the antioxidant was assumed to oppose the reaction of oxygen with the polymer. The term "anti-oxygen", an outgrowth of this theory, was used for many years to describe those additives that inhibited oxidation.

A chain-reaction mechanism had been proposed by Bodenstein [9] in 1913 to explain the photochemical synthesis of hydrogen chloride. Several investigators [10-12] adapted this chain-reaction concept to the autoxidation of hydrocarbons and hydrocarbon polymers. Although advances could be made in the development of antioxidants by the trial-and-error method, the growth in number and variety of stabilizers available today required at least a partial understanding of the complex reactions responsible for autoxidation. Once these mechanisms were established, a scientific approach to stabilization was possible. As the chain-reaction mechanism received general acceptance, the term, "anti-oxygen" was abandoned in favor of the current term, "antioxidant". As will become evident later, antioxidants are but one type of stabilizer, used primarily to inhibit thermal oxidation. Other types of stabilizers have been developed to protect polymers against photooxidation and ozone-induced degradation.

I) Hydrocarbon Polymers

The earliest mechanisms accounting for stabilization were developed in studies on the autoxidation of rubber, and eventually these were adapted to the synthetic hydrocarbon polymers. It is logical, therefore to explore these mechanisms before considering the stabilization of more complex polymers. However, much that has been learned about mechanisms for the stabilization of hydrocarbon polymers serves as a background for understanding stabilization reactions in more complex polymers. The free-radical-initiated, chain-reaction mechanism developed to explain thermal oxidation is similar to reactions which take place during photooxidation, but there are some important differences between the two mechanisms.

1) Short-term and Long-term Antioxidants

Polymers are normally processed in the molten state and thus are exposed to elevated temperatures albeit for only short intervals. As would be expected, the rate of oxidation increases with temperature, and excessive oxidation is quite likely to occur under these conditions – unless the polymer has been adequately stabilized. Sites of incipient degradation, formed during processing, lead to eventual failure under conditions of normal use. Short-term or processing antioxidants are designed to provide protection during processing or fabrication into finished products. For exceptionally sensitive polymers, it may even be necessary to provide protection at the final stage of polymerization.

Polypropylene undergoes extensive degradation under commercial processing conditions [13]. Hydroperoxides form at a significant rate, and these reactive intermediates decompose into radicals capable of promoting either thermal or photooxidation. A variety of antioxidants has been used to inhibit degradation at this critical stage in the life cycle of thermally-sensitive polymers [14]. Low-molecular-weight phenols such as 2,6-ditert. butyl-4-methylphenol,

$$OH$$
$$(CH_3)_3C \quad C(CH_3)_3$$
$$CH_3$$

are effective short-term antioxidants. The diethyldithiocarbamates of zinc and nickel have also been found to be effective [14], particularly as antioxidants in natural rubber.

Antioxidants intended to provide protection during processing must be capable of migrating freely throughout the polymer mass to reach the large number of initiation sites that are generated at elevated temperatures. For this reason, low-molecular weight antioxidants are preferred. Short-term antioxidants, as the term implies, are not intended to give protection during extended use. For protection under these conditions, long-term antioxidants must be used.

In contrast to processing antioxidants, those stabilizers designed for long-term protection are usually large, complex molecules which have less mobility through the polymer but are also less volatile and so have superior retention. These antioxidants may function either as preventive or as chain-breaking stabilizers although many function by both mechanisms. The kinetic scheme for autoxidation of hydrocarbon polymers (refer to Sect. B) shows the stages in degradation at which each type of antioxidant is effective. The basic properties of antioxidants are contrasted in Table D-1.

D) Stabilization Against Thermal Oxidation

Table D-1. Basic properties of antioxidants

Short-term antioxidants	Long-term antioxidants
Low Molecular Weight	High Molecular Weight
High Mobility	Low Mobility
Retention not Necessary	High Level of Retention

Initiation:

$$ROOH \rightarrow RO^{\bullet} + HO^{\bullet}$$

or

$$2\,ROOH \rightarrow RO^{\bullet} + ROO^{\bullet} + H_2O.$$

Propagation:

$$ROO^{\bullet} + RH \rightarrow ROOH + R^{\bullet}$$

and

$$R^{\bullet} + O_2 \rightarrow ROO^{\bullet}.$$

Oxidative Chain Branching

$$ROOH \rightarrow RO^{\bullet} + HO^{\bullet}$$

$$RO^{\bullet} + RH \rightarrow ROH + R^{\bullet}.$$

In the initiation step, preventive antioxidants cause hydroperoxides to decompose by an alternate mechanism which does not yield radicals capable or propagating oxidation. If transition metals are present, these catalyze the decompositon of hydroperoxides into radicals, and another type of preventive antioxidant must be used to deactivate the metal.

Chain-breaking antioxidants are effective in the propagation stage of autoxidation. They compete with the polymer to trap or react with those radicals which propagate oxidation thus reducing the length of the oxidative chain. This reaction is represented by stabilization with a labile-hydrogen donating antioxidant (HA),

$$ROO^{\bullet} + HA \rightarrow ROOH + A^{\bullet}.$$

The byproduct of this reaction, $A^{\bullet}$, though it is a free radical, does not propagate oxidation as do the polymer radicals, $R^{\bullet}$, $RO^{\bullet}$, $ROO^{\bullet}$ etc. These $A^{\bullet}$ radicals may trap and terminate a second propagating radical, or they may be deactivated by coupling or disproportionation.

For many applications, antioxidants must also be nonstaining when added to the polymer and should not form chromophores under reaction conditions.

2) Chain-breaking Antioxidants

In the chain reaction of uninhibited oxidation, as many as a hundred steps of propagation may result from a single initiation event [15]. Despite the fact that this reaction is responsible for the rapid degradation of hydrocarbon polymers, it presents an important step at which stabilization could be realized by antioxidants that interrupt or shorten the oxidative chain. Several propagation steps may occur before the chain is interrupted, but a reduction from hundreds of steps to only a few will represent an appreciable degree of stabilization.

Chain-breaking antioxidants were historically the first type of stabilizers to be used in the protection of polymers against autoxidation. Two types of chain-breaking antioxidants have been found to be effective, each functioning by a different mechanism.

a) Free-Radical Traps

The chain reaction responsible for autoxidation is initiated by free radicals, and hence reactions which deactivate or reduce the reactivity of initiating radicals will contribute to stability. Scavenging or trapping of radicals responsible for degradation is one obvious approach. Radical trapping has been used to inhibit polymerization. For example, the spontaneous polymerization of styrene is inhibited by addition of small amounts of quinone[16]. One mechanism suggested to account for this inhibition is based on addition of styryl radicals to the aromatic ring,

The radical (I), formed from the inhibitor, is relatively stable and hence unlikely to react rapidly enough with styrene to continue the propagation of kinetic chains. Also, trace impurities such as hydroperoxides are believed to be initiators for styrene polymerization[17]. Trapping of radicals from decomposing peroxides could also inhibit spontaneous polymerization.

To be effective in the stabilization of polymers, radical traps should be unreactive toward the host polymer. Relatively stable nitroxide radicals are capable of trapping propagating radicals[18]. These include among others, di-tert-butyl nitroxide (II)

and 2,2',6,6'-tetramethyl-4-pyridone nitroxide (III).

Although relatively stable, nitroxide radicals are not stable enough to permit extensive commercial application. The diaryl nitroxides are more stable but even these are not considered to be practical stabilizers.

Although radical traps are of scientific interest for reaction mechanism studies, they are in general of less interest than the more important class of commercial antioxidants which inhibit oxidation by donating a labile hydrogen to deactivate propagating radicals. Carbon black, which can function as a thermal antioxidant, could act as a radical trap, but the weight of evidence indicates that it is a labile-hydrogen donor. This unusual stabilizer, used primarily to inhibit photooxidation is discussed in Sect. D-I-7.

b) Labile-hydrogen Donors

Antioxidants which contain one or more labile hydrogens in their structure are the most frequently used stabilizers for protecting polymers against autoxidation. Many important antioxidants for both short and long-term protection function by this mechanism. These hydrogen-donating antioxidants (HA) provide an alternative reaction to the rate-controlling step in the propagation phase of autoxidation.

$$ROO^\bullet \begin{cases} \xrightarrow{RH} ROOH + R^\bullet & \text{(propagation)} \\ \xrightarrow{HA} ROOH + A^\bullet & \text{(stabilization)} \end{cases}$$

The key of their effectiveness is the greater ease of hydrogen removal in comparison to abstraction of hydrogen from polymer molecules [19]. The direct reaction of a labile-hydrogen donor with oxygen must be restricted, however, to minimize loss of hydrogen before free-radical, terminating reactions can take place.

The hydrogen abstraction mechanism has been questioned by Hammond [20, 21] who failed to find a kinetic isotope effect when deuterium was substituted for the labile hydrogen in several antioxidants. The higher bond dissociation energy of the C—D bond as compared with that of the C—H bond would be expected to reduce the effectiveness of antioxidants modified in this way. However, Hammond and his associates did not find this anticipated isotope effect. They suggested as an alternative mechanism the formation of an unstable complex as an intermediate in the reaction,

$$ROO^\bullet + HA \rightarrow [ROO\text{—}HA]$$
$$[ROO\text{—}HA] + ROO^\bullet \rightarrow \text{inactive products}$$

There was no direct evidence for the proposed intermediate, which was not isolated, although indirect evidence for its existence was obtained by other investigators [22, 23].

Through a series of carefully controlled experiments, Shelton and coworkers [24, 25] eventually demonstrated that there is indeed an isotope effect with deuterated antioxidants. A key point in Shelton's work was recognition that either the normal or deuterated antioxidants also function as initiators by their direct oxidation to yield A$^\bullet$ radicals.

$$HA + O_2 \rightarrow HOO^\bullet + A^\bullet .$$

This reaction is now considered to be an important initiation reaction, particularly in elastomers which are protected with labile-hydrogen donors. The loss of deuterated antioxidants during initiation would mask the isotope effect and make it very difficult

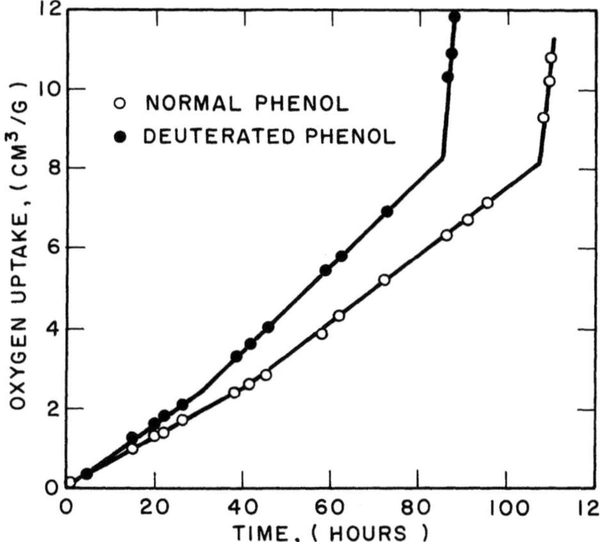

Fig. D-1. Oxidation of purified polyisoprene containing I PPH 2,6-DI-T-butyl-4-methylphenol (90°C, I ATM O$_2$). (Reprinted with permission of The American Chemical Society)

to detect. Shelton was able to overcome this problem by using a higher concentration of the antioxidant and observing the reaction at higher temperatures where participation of the antioxidant in initiation is reduced. There is also the possibility of deuterium exchange with hydrogen from water formed during oxidative degradation.

An example of the isotope effect observed by Shelton is shown in Fig. D-1. In the inhibited oxidation of highly purified polyisoprene, two different rates are evident in the retarded stage of the reaction. This change in rate, observed after about 40 h, is taken as support for the role of antioxidants in initiation. The direct oxidation of the antioxidant is believed responsible for the first slow rate. Hydroperoxide decomposition then increases during the second stage of inhibited oxidation, and radicals from this source initiate new oxidative chains accounting for the more rapid rate. Similar isotope effects have since been demonstrated [26, 27] in other polymers and with both deuterated amine and deuterated phenolic antioxidants.

Both secondary amines and hindered phenols function as labile-hydrogen doners in protecting hydrocarbon polymers and elastomers against autoxidation. The secondary amines were first used in the stabilization of natural rubber. The bond between hydrogen and nitrogen has a dissociation energy of 93 kcal/mol in contrast to 95 kcal/mol for C—H bonds of methylene groups in the backbone chain of rubber molecules. Also the smaller antioxidant molecules have much higher mobility and so are more likely to intercept radicals than the less mobile polymer molecules. Reaction with other hydrogens in the structure is even less favored since these C—H bonds have significantly higher dissociation energies. Thus the antioxidant is able to compete effectively for reaction with propagating radicals. Secondary amines also protect hydrocarbon polymers but are used less frequently in synthetic polymers.

N,N'-diphenyl-p-phenylenediamine (IV) and phenyl-β-naphthylamine (V) are typical secondary amine antioxidants that have been used for many years to inhibit the autoxidation of rubber. Most amine antioxidants are discolored or produce color in rubber stocks as oxidation progresses. The reaction of diphenyl-p-phenylenediamine is

45

an illustration of how color can develop during an inhibited autoxidation. This amine contains two labile hydrogens, each of which can interrupt an oxidation chain.

The resulting diimine is an intense chromophore, accounting for the discoloration produced by this antioxidant. N,N'-dinaphthyl-p-phenylenediamine in contrast produces very little discoloration and is often described as a nonstaining amine antioxidant. Recent evidence suggests that certain of the amines previously used as antioxidants are carcinogens[28]. Most of these have now been removed from commercial use.

Phenolic antioxidants were developed to protect synthetic hydrocarbon polymers where discoloration often cannot be tolerated. Discoloration is not as important in rubber formulations which are usually pigmented with carbon black. The structure of phenolic antioxidants emphasizes the importance of restricting direct oxidation of the antioxidant. The labile hydrogen which is on the hydroxyl group is shielded with bulky groups (R) at one or both ortho positions.

Phenolic antioxidants of many varying structures are now available commercially. Improved retention and greater efficiency have been important factors in this proliferation of phenolic antioxidants. Table D-2 lists several of the important phenolic antioxidants designed for long-term protection.

Reactions by which phenolic antioxidants inhibit autoxidation have been established for some of these stabilizers. For example, the short-term antioxidant, 2,6-ditert. butyl-phenol, is capable of interrupting two oxidative chains although this compound contains only one labile hydrogen[29],

Table D-2. Typical phenolic antioxidant structures

Simple Phenols:

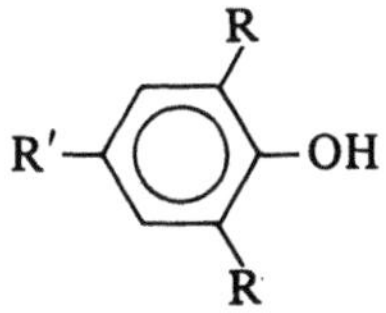

Methylenebisphenols:

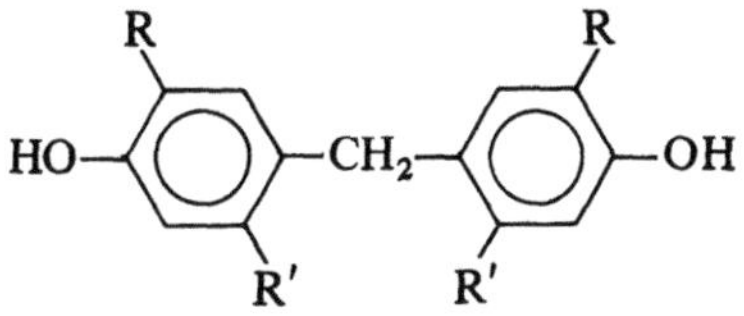

Thiobisphenols:

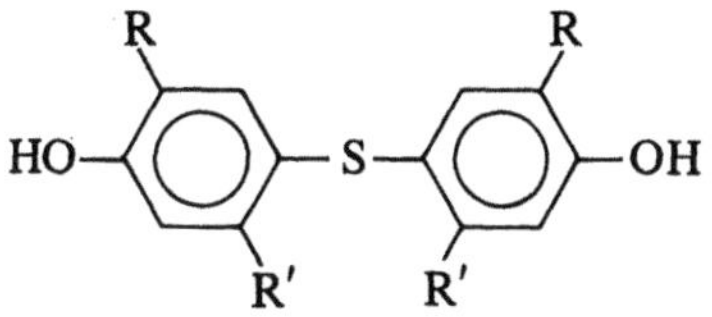

Polyhydric Phenols:

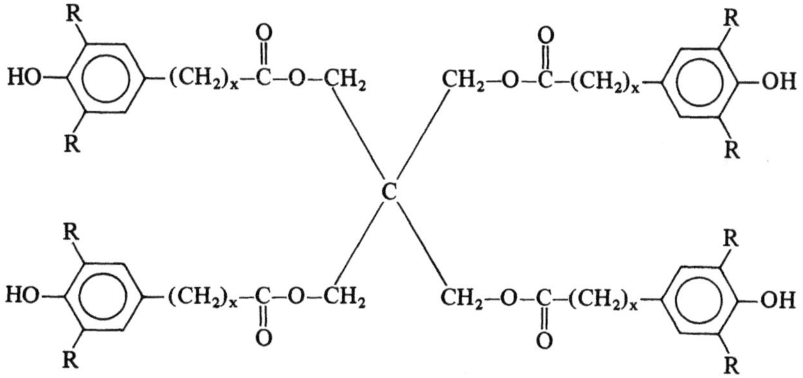

Where R is a bulky group such as tert. butyl and R' is methyl.

Long-term phenolic antioxidants shown in Table D-2 have from one to four labile hydrogens available for interrupting oxidation chains. These antioxidants have the further advantage of higher molecular weight which results in increased retention. Loss of antioxidants by extraction or leaching can be important in certain applications. Selection of an antioxidant for such an application must take into account solvents which will be in contact with the inhibited polymer.

3) **Preventive Antioxidants**

Whereas chain-breaking antioxidants inhibit at the propagation stage of autoxidation, preventive antioxidants stabilize by reducing the rate of initiation. The dithiocarbamates were known to inhibit the oxidation of natural rubber before their preventative function was recognized. It was not until the important role of hydroperoxides in oxidation was fully understood [2, 3] that the mechanism by which these important rubber stabilizers function was established [30, 31].

Support for the preventative action of antioxidants is found in the work of Kennerly and Paterson [32] who showed that a wide variety of additives, including phenols and certain compounds of sulfur or phosphorous, decompose hydroperoxides into nonradical products. Thus there are two different mechanisms for the decompositon of hydroperoxides formed during oxidation. The first of these, the normal homolysis into radicals, is accelerated by heat and catalyzed by active metals. The second is an induced decomposition which probably proceeds by ionic reactions. Whatever the mechanism, radicals capable of propagating oxidation are not formed by this second mechanism. These alternate reactions had been reported previously by Kharasch [33] who observed both mechanisms in the decomposition of cumene hydroperoxide.

Cumene hydroperoxide has been a very useful model compound for studying hydroperoxide reactions in inhibited autoxidation. The products of its decomposition are quite different depending on the decomposition mechanism, and are easily identified or quantified by standard analytical procedures. Apparently the induced decomposition is catalyzed by acidic compounds. Similar reactions have been used to explain inhibition of autoxidation in other hydrocarbon polymers by preventative antioxidants.

There are other types of preventative antioxidants which inhibit by supressing the catalytic effect of active metals that promote hydroperoxide decomposition into radicals. These preventative antioxidants are referred to as metal deactivators, and will be discussed in a following section.

a) Hydroperoxide Decomposers

The reactions involved in stabilization with sulfur compounds have been studied [34–36] more extensively than those for other hydroperoxide decomposers. Alkyl sulfides and aryl disulfides are effective inhibitors for the oxidation of natural rubber and synthetic hydrocarbon polymers respectively. Both apparently react with hydroperoxide to yield nonradical products. The mechanisms for inhibition by these additives, however, are quite complex, involving the following sequence of reactions [34].

Alkyl Sulfides,

$$(CH_3)_3CSC(CH_3)_3 \xrightarrow{ROOH} (CH_3)_3C\overset{\overset{\textstyle O}{\uparrow}}{S}C(CH_3)_3$$

$$(CH_3)_3C\overset{\overset{\textstyle O}{\uparrow}}{S}C(CH_3)_3 \xrightarrow{65-100\ °C} (CH_3)_3CSOH + (CH_3)_2C{=}CH_2$$

tert-Butanesulfenic Acid

$$2(CH_3)_3CSOH \xrightarrow{-H_2O} (CH_3)_3C\overset{\overset{\textstyle O}{\uparrow}}{S}SC(CH_3)_3$$

tert-butyl-tert-butane-thiosulfinate

$$(CH_3)_3CSSC(CH_3)_3 \xrightarrow{ROOH} (CH_3)_3C\overset{\overset{\textstyle O}{\uparrow}}{S}SC(CH_3)_3$$

$$(CH_3)_3C\overset{\overset{\textstyle O}{\uparrow}}{S}SC(CH_3)_3 \xrightarrow{heat} (CH_3)_2C{=}CH_2 + (CH_3)_3CSSOH$$

tert-butanethiosulfoxylic acid

Addition of $CaCo_3$ to a benzene solution of cumene hydroperoxide containing ditert.-butyl sulfoxide reduced the catalytic effect of the sulfur compound in hydroperoxide decomposition. Addition of $CaCO_3$ to a similar reaction mixture which did not contain the sulfoxide had no effect on hydroperoxide decomposition. Thus it is evident that acidic products are the actual hydroperoxide decomposers as shown in Fig. D-2.

The sulfenic acid in the preceding reaction sequence is believed to be a primary hydroperoxide decomposer. In benzene solution it reacts rapidly to decompose two moles of hydroperoxide. The thiolsulfinate, formed from the disulfide, appears to be even more effective, but the most active product is believed to be the corresponding thiosulfoxylic acid or other acidic decomposition products. After a slow, initial reaction, during which the thiosulfoxylic acid is formed, there follows a catalytic reaction in which many moles of hydroperoxide are destroyed per mole of thiolsulfinate added. In each case, the products of cumene hydroperoxide decomposition were phenol and acetone as expected from an acid-catalyzed reaction. Thus sulfides and disulfides are not the preventative antioxidants but function as precursors for the actual hydroperoxide decomposers.

D) Stabilization Against Thermal Oxidation

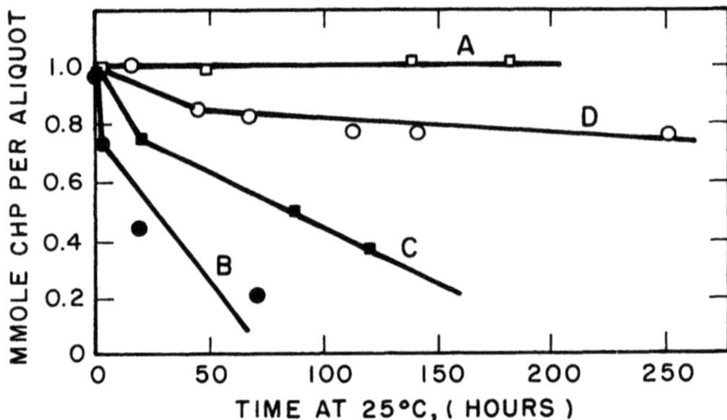

Fig. D-2. Effect of base on decomposition of cumene hydroperoxide in the presence of tert-butanesulfenic acid. Concentrations, mmol/l in benzene: CHP 2.0 ± 0.1; A, □ $CaCO_3$ 0.08; B, ● tert.-BuSOH 0.2; C, ■ tert.-BiSOH 0.2, $CaCO_3$ 0.08; D, ○ tert.-BuSOH 0.2, $CaCO_3$ 2.5 (15). (Reprinted with permission of Rubber Chemistry and technology)

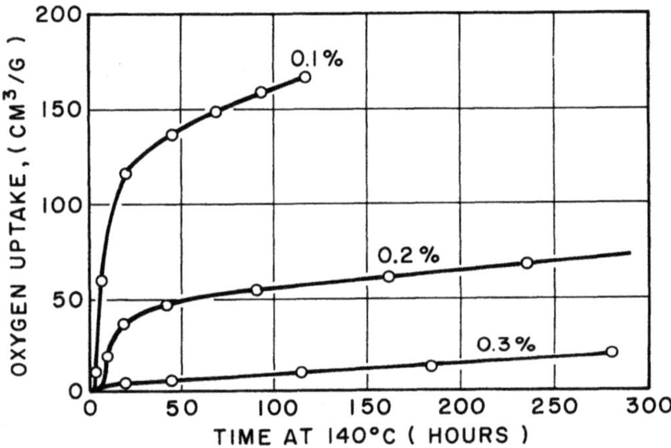

Fig. D-3. The effect of varying concentrations of naphthyl disulfide on the stabilization of low-density polyethylene. (Reprinted with permission of John Wiley and Sons, Inc.)

Hawkins and coworkers [37, 38] have investigated the action of aryl disulfides as hydroperoxide decomposers in the autoxidation of polyethylene. Naphthyl disulfide at relatively high concentration as shown in Fig. D-3 inhibits oxidation at 140 °C. Concentrations of 0.1 and 0.2% have little effect initially, but as the reaction proceeds the oxidation rate is reduced sharply. At 3% concentration the inhibited polymer oxidizes slowly from the start with no indication of an autocatalytic reaction. As with the alkyl sulfides, naphthyl disulfide reacts with hydroperoxides formed in the polymer, yielding the corresponding thiolsulfinate,

$$C_{10}H_7-SS-C_{10}H_7 \xrightarrow{ROOH} C_{10}H_7-\overset{\overset{\displaystyle O}{\uparrow}}{S}S-C_{10}H_7 \, .$$

Although the reaction sequence beyond the thiosulfinate has not been established as was done with alkyl sulfides, evidence has been obtained showing that either the thiol-

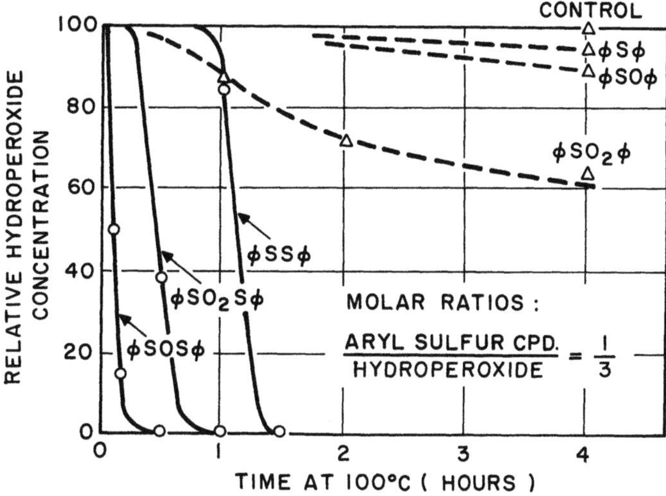

Fig. D-4. The effect of aryl sulfur compounds on the decomposition rate of cumene hydroperoxide (5%) in cumene. (Reprinted with permission of The Sciety of Plastics Engineers)

sulfinate or acidic products formed by its rapid thermal decomposition are the actual hydroperoxide decomposers. This is apparent in Fig. D-4 which shows decomposition rates for a 2% solution of cumene hydroperoxide in cumene at 90 °C. Addition of naphthyl disulfide has only a small effect on the decomposition rate for the first hour. Then a rapid decomposition occurs with complete decomposition of the hydroperoxide. When the thiolsulfinate is added rather than the disulfide, decomposition takes place immediately. The disulfide reacts stoichiometrically with hydroperoxide to yield the thiolsulfinate and this accounts for the observed induction period. Once the concentration of thiolsulfinate has reached a critical level, rapid hydroperoxide decomposition occurs.

Sulfur dioxide was identified [37] as one of the products formed in the decomposition of dinaphthyl disulfide. This acidic product is more effective by orders of magnitude than the thiolsulfinate. Sulfur dioxide would have only a transitory existence in the polymer but this very active product would form at those sites where hydroperoxides are forming. These results closely parallel those of Shelton [34], showing again the role of acidic products in hydroperoxide decomposition. It is interesting to note that Kharasch [33] used SO_2 in his study of the induced decomposition of cumene hydroperoxide.

Alkylated thiobisphenols have been used for many years as antioxidants for polyethylene and polypropylene. Since these additives contain labile hydrogens, they should act as chain-breaking antioxidants. They could, however, also function as preventative antioxidants through reactions at the sulfur bridge. Chasar [39] has compared the effectiveness of thiobisphenols with their oxidation products, the sulfoxides and sulfones. This study was conducted in polypropylene rather than model compounds. Hydroperoxide decomposition was not observed directly, but it was concluded that both the thiobisphenols and their corresponding sulfoxides decomposed during processing to yield traces of sulfur dioxide. The sulfoxide was found to be more effective than the corresponding sulfide, and it was proposed that this was a result if its greater instability to heat. The sulfone was virtually ineffective, as might be expected from its greater thermal stability.

Metal salts of dialkyldithiophosphates have been used as antioxidants in rubber for many years. They are of particular interest in that they apparently function simultaneously as chain-breaking and preventative antioxidants [40]. As is evident in the structure of the zinc salt of diethyl dithiocarbamate (VI), these compounds do not contain the labile hydrogens similar to those existing in secondary amines and hindered phenols. However, an electron-transfer process has been proposed [40, 41] to account for the deactivation of propagating radicals, by these metallic salts.

$$\left[\left(\begin{array}{c}CH_3 \\ CH \\ CH_3\end{array}\right)_2 PS_2\right]_2 Zn \ + \ 2\,ROO^{\bullet}$$

VI

$$\longrightarrow \left(\begin{array}{c}CH_3 \\ CH \\ CH_3\end{array}\right)_2 P \begin{array}{c}S-S \\ \\ S \ \ S\end{array} P-O\left(HC\begin{array}{c}CH_3 \\ \\ CH_3\end{array}\right)_2 \ + \ Zn^{++} \ + \ 2\,ROO^{-}$$

There is also evidence [42] that hydrogen may be abstracted from the alkyl groups in these compounds.

$$\left[\left(\begin{array}{c}CH_3 \\ CH \\ CH_3\end{array}\right)PS_2\right]_2 Zn \ + \ 2\,ROO^{\bullet} \longrightarrow \left[\left(\begin{array}{c}CH_3 \\ C^{\bullet} \\ CH_3\end{array}\right)PS_2\right]_2 Zn \ + \ 2\,ROOH$$

The role of dithiophosphates in hydroperoxide decomposition also has considerable support [32, 43]. Obviously the mechanism by which these additives inhibit oxidation is quite complex. It is further complicated by observations [44] that they can also contribute to initiation by interaction with hydroperoxides to form free radicals.

Reactions of the dithiophosphates serve to point out an important generality in antioxidant reactions. Although most antioxidants inhibit primarily in either chain-breaking or perventative reactions, often they perform both functions with one the principal and the other a minor role.

b) Metal Deactivators

Homolytic decomposition of hydroperoxides is catalyzed by traces of metal ions which are capable of undergoing one-electron, oxidation-reduction reactions [45–48]

$$ROOH + M^{n+} \rightarrow RO^{\bullet} + M^{(n+1)^+} + HO^-$$

and

$$ROOH + M^{(n+1)} \rightarrow ROO^{\bullet} + M^{n+} + H^+.$$

Those ions which are strong reducing agents, e.g. ferrous ions, react to form alkoxy radicals whereas oxidizing ions, e.g. lead ions, generate peroxy radicals as the major product. If the potential for oxidizing and reducing are approximately equal, as with

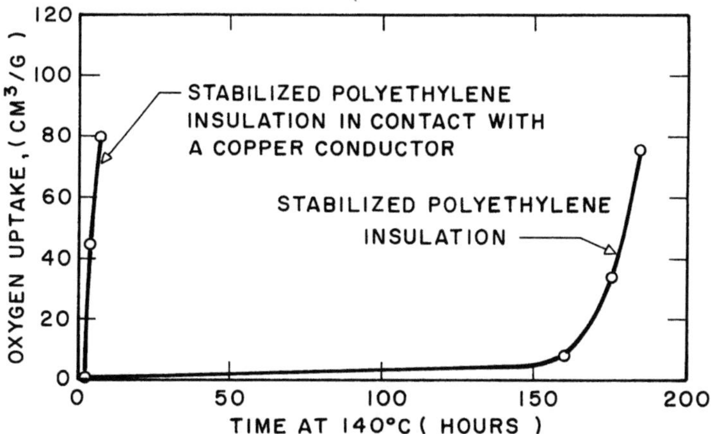

Fig. D-5. The effect of copper on the oxidative stability of black PE insulation. (Reprinted with permission of The Society of Plastics Engineers)

cobalt ions, both reactions occur simultaneously with rapid degradation of the polymer. When these reactive metals or their ions are in contact with a stabilized polymer, oxidation takes place almost as rapidly as in the uninhibited polymer. Polypropylene, in which tertiary hydroperoxides predominate, is much more susceptible to metal-catalyzed oxidation than is polyethylene.

Preventative antioxidants, introduced into polymers to induce the decomposition of hydroperoxides into nonradical products, cannot compete with metal catalysis. This is shown in Fig. D-5. Oxidation of a low-density polyethylene insulation is inhibited for over 150 h at 140 °C. This same composition when extruded over a copper conductor, however, loses almost all of its stability. Metal-catalyzed oxidation is responsible for the rapid degradation of polymers laminated to active metals or when used as wire insulation. Fortunately, stabilizers are now available to offset, at least in part, the destructive effects of metal catalysis.

Hansen and coworkers [48] found that oxamide and several of its derivatives, e.g. oxanilide (VII), function as metal deactivators. Their effectiveness in protecting against metal-catalyzed oxidation is attributed to their ability to form coordinated complexes with the active transition metals. Metal deactivators that are available today do not completely eliminate metal catalysis.

$$C_6H_5NHCOCONHC_6H_5 \quad VII$$

This is seen in Fig. D-6, which shows the effect of oxanilide on the oxidation of polyethylene stabilized with 4,4'-thiobis(2-tert. butyl-5-methylphenol) and in contact with copper. Oxidation of the stabilized formulation is inhibited for about 600 h, but this protection is completely lost on contact with copper. Addition of increasing amounts of oxanilide up to 0.5 percent restores only a fraction of the original stability.

Allara and Chan [50] have synthesized typical complexes of copper with oxamide derivatives by the following reaction,

$$Cu(OAc)_2 + 2\,RNHCOCONHR \rightarrow Cu[RNCOCONR]_2 + 2\,HOAc.$$

(In oxanilide, R would be C_6H_5)

D) Stabilization Against Thermal Oxidation

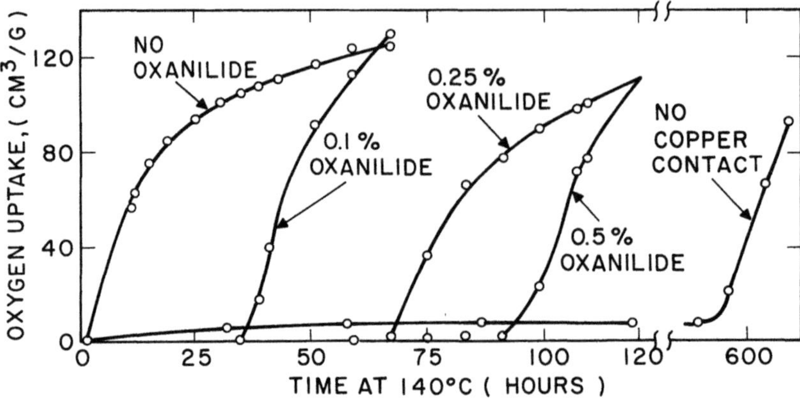

Fig. D-6. The effect of oxanilide on the oxidative stability of polyethylene containing 0.1% 4,4'-thiobis(2-tert. butyl-5-methylphenol) and in contact with copper. (Reprinted with permission of The American Chemical Society)

Polymeric structures (VIII) and (IX) have been proposed for these complexes on the basis of elemental analysis and infrared spectroscopy. When these complexes are

VIII

IX

blended into polyethylene, they also function as oxidation catalysts, but to a lesser extent than copper or its oxides, as shown in Table D-3 of this part. Although complete

Table D-3. Relative specific rate constants for maximum oxidation of Polyethylene at 100 °C when catalyzed by copper oxides and copper complexes

Copper compound	Relative specific rate constants[a]
Cu_2O	19.6
CuO	1.3
Cu(oxamide)	0.52
Cu(oxanilide)	1.1
Cu(N,N'-dibenzaloxalyl dihydrazide)	1.0

[a] As a function of catalyst surface area

54

deactivation of the metal does not take place, addition of metal deactivators to stabilized polymers does provide adequate protection in most applications.

Copper catalysis of polymer oxidation takes place initially at the interface between metal and polymer in laminates and in insulated wire. Prior blending of copper carboxylates into polyethylene also catalyzes oxidation. Similar salts are also formed at the interface as degradation products formed in the matrix migrate to the metal interface. Inhibition by metal deactivators could then take place either at the interface, or within the matrix as these salts migrate through the polymer. Determination of the primary process for stabilization is important to the future development of improved metal deactivators.

Allara and White [51] have examined the rates of migration of copper salts through low-density polyethylene. Copper exchange between these salts and metal deactivators would result in stabilization within the matrix. Conversely, the inhibitor may first migrate to the interface and deactivation would then occur at this point. Addition of a series of copper salts of alkyl carboxylates, $Cu(C_nH_{2n-1}CO_2)_2$ in which n = 3, 7, 11, 17, and 29, believed to be typical of products formed in the oxidizing polymer, increased the rate of oxidation, but oxanilide had little effect on the rate at which these salts diffused through the polymer or on stabilization. Based on their observations, these authors concluded that surface or interface reactions are the primary inhibition process rather than copper exchange reactions between the stabilizer and carboxylic degradation products within the matrix. However, as oxidation proceeds and the concentration of copper salt increases, stabilization reactions in the polymer bulk probably do contribute to stabilization in a secondary process. Similar processes probably take place in oxidation catalyzed by other active metals.

4) Antioxidant Combinations

Often it is possible to obtain a high level of protection by using combinations of two or more different stabilizers. Combinations of antioxidants with photostabilizers, which will be discussed later, are very effective in protecting polymers against the combined effects of thermal and photooxidation. In thermal oxidation, the proper selection of antioxidants can provide very high levels of protection. Combinations of antioxidants may result in (1) a simple additive effect, (2) antagonism, or (3) synergism.

a) Additive Effects

Usually when two or more antioxidants of the chain-breaking type are used in combination, a simple additive effect results. This effect, however, may exceed that obtained when the concentration of either component is increased twofold. There is a limit to the amount of an antioxidant that is retained by a polymer. Two or more chain-breaking antioxidants could provide better retention than higher concentrations of either component. A higher level of protection over an extended temperature range is also possible. Components of such a combination would be selected for their efficiency over a portion of the temperature range to which the polymer will be exposed. The combination of a short-term or processing antioxidant with one designed for long-term protection would be an example of such a combination.

b) Antagonistic Effects

Antagonism between antioxidants would be expected if interactions occur which would destroy or reduce the effectiveness of either component. Hawkins and coworkers [52] observed an antagonistic effect when carbon black, used to protect against outdoor weathering, was combined with secondary amines or certain hindered phenols. Figure D-7 shows the effect of combining carbon black and N,N'-diphenyl-p-phenylene diamine in polyethylene.

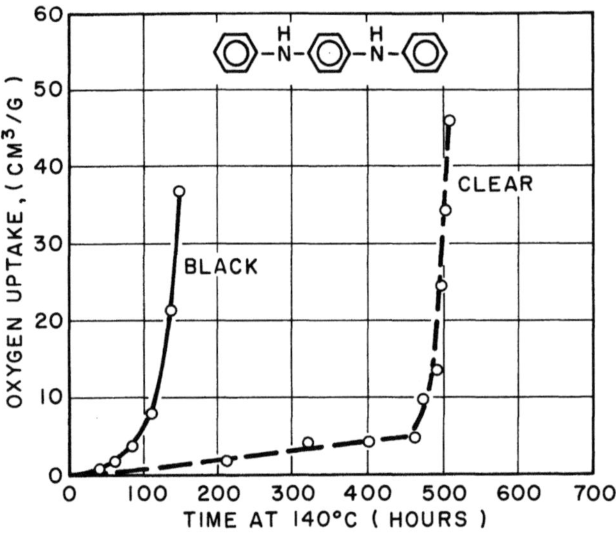

Fig. D-7. Effect of carbon black in 3% conc. on the thermal oxidation of polyethylene containing 0.1% N,N'-diphenyl-p-phenylenediamine

Carbon black is a weak thermal antioxidant for polyethylene at 140 °C, inhibiting oxidation for only about 20 h when used alone. The amine, on the other hand, provides effective protection for over 450 h. When the two are used in combination, however, the polymer is protected for only about 100 h. Carbon black also reacts antagonistically with many hindered phenols.

Absorption of amines or phenols onto the carbon black surface has been suggested as an explanation for the antagonistic effect. However, the extent of the antagonism varies with the chemical structure of carbon blacks, suggesting that catalytic destruction of the antioxidant at the surface may be the major factor responsible for antagonism. There are certain phenols which exhibit synergism with carbon black rather than antagonism.

c) Synergism

When combinations of two or more antioxidants provide more protection than would be expected from the sum of that provided by the individual components, the phenomenon is referred to as synergism. Synergistic affects have been observed [27] with combinations of several different types of antioxidants, and many of these combinations are now used commercially. When a synergistic combination consists of two or more antioxidants, each functioning by the same mechanism but of unequal activity, the phenomenon has been referred to as homosynergism [53]. The more common type, heterosynergism, involves two or more antioxidants acting at different steps in the degradation mechanism.

Homosynergism has been observed [54] with combinations of two hindered phenols, acting as chain-breaking antioxidants. In these combinations, the more effective phenol (X) is the primary deactivator of propagating radicals. The less active phenol (XI) functions as a reservoir for labile hydrogen to regenerate the primary antioxidant. In this example, the two phenols differ in the size and number of the ortho substituents which affects the ease of hydrogen abstraction. Phenol X has less hindrance adjacent to the labile hydrogen and is therefore more effective. Since it is regenerated from its radical by abstraction of hydrogen from XI, it is not depleted as rapidly. Combinations of amines and of amines with phenols are also homosynergistic, presumably by regeneration of the more effective component [54].

Heterosynergistic combinations are often necessary to provide adequate protection for readily-oxidized hydrocarbon polymers, or when long service life is a requirement. Most combinations of this type consist of a preventive and a chain-breaking antioxidant. One component in these combinations may be only a weak stabilizer when used alone but yet function as an effective synergist.

Carbon black is an excellent stabilizer against photooxidation and it also functions as a mild thermal antioxidant, trapping or scavenging propagating radicals. Under accelerated test conditions, it inhibits the oxidation of polyethylene for short periods. When used in combination with thiobis-2-naphthol, however, it is an effective synergist in polyethylene as shown in Fig. D-8. The role of carbon black in this synergistic com-

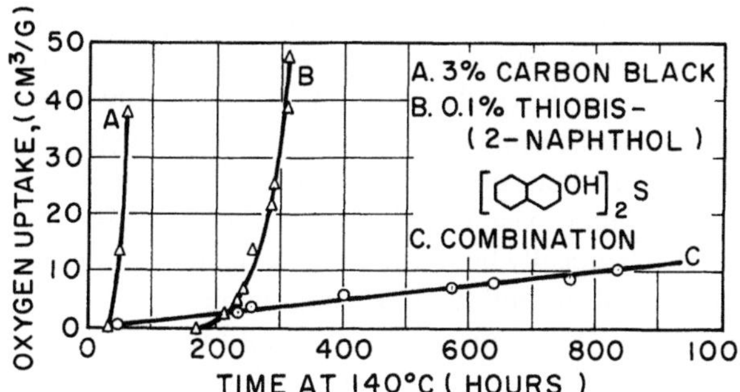

Fig. D-8. Stabilization of polyethylene with a synergistic combination consisting of 0.1% thiobis-(2-naphthol) and 3% carbon black

bination is probably that of a hydroperoxide decomposer although it may also contribute as a chain-breaking antioxidant. Several other sulfur-containing antioxidants show heterosynergism with carbon black and with certain polyacenes [55–57].

Many heterosynergistic combinations have been reported [57, 58] that do not involve carbon black. Synergism has been observed [58] with combinations of hindered phenols and dialkylphosphonates or thioethers [57]. Combinations of hindered phenols and thioethers such as dilauryl thiodipropionate (XII)

$$(C_{12}H_{25}O-\overset{\overset{\displaystyle O}{\|}}{C}-CH_2-CH_2)_2S \quad (XII)$$

are widely used to protect polypropylene against oxidative degradation.

Although regenerative mechanisms can account for synergism when hydrogen transfer can take place between the two antioxidants in the combination, in many instances this mechanism is inadequate to explain the phenomenon. There is, however, an alternative mechanism for synergism. Chain-breaking antioxidants shorten the length of oxidative chains, but a molecule of hydroperoxide is formed for each termination step,

$$ROO^{\bullet} + HA \rightarrow ROOH + A^{\bullet}.$$

The concentration of hydroperoxides is greatly reduced, however, so that there is less of it to be deactivated by the hydroperoxide decomposer. This is equivalent to an increase in the initial concentration of the preventative antioxidant. (The effect of increasing concentration of a preventative antioxidant was shown in Fig. D-3.) Reduction in the concentration of radicals formed by hydroperoxide decomposition then conserves the chain-breaking antioxidant since it would be consumed by the following reactions.

$$2\,ROOH \rightarrow ROO^{\bullet} + ROO^{\bullet} + H_2O$$

$$ROO^{\bullet}(R^{\bullet}, RO^{\bullet} etc) + HA \rightarrow ROOH\,(RH, ROH) + A^{\bullet}.$$

Thus each antioxidant in these combinations protects and preserves the other. This is probably the predominant mechanism responsible for heterosynergism with many antioxidant combinations.

5) Nonmigrating Antioxidants

Retention of antioxidants is essential when polymers must resist oxidative degradation for long periods of time. Even the high-molecular-weight antioxidants migrate, albeit slowly, and have a vapor pressure sufficient to cause loss by evaporation. Although the solubility of antioxidants in water is very low, it is still sufficient to slowly dissolve many of these additives. Antioxidants may also be removed from the surface by wiping or abrasion. Completely nonmigrating antioxidants provide the ultimate in retention.

Antioxidants will be nonmigrating if they exist in the polymer as large, bulky particles, e.g. carbon black, or if they are chemically bound to polymer chains. Inhibition by these immobile species must depend on the ability of propagating radicals to move through the polymer, without reacting with polymer molecules, until they encounter the immobile stabilizer. This process in unlikely to occur at elevated temperatures, but nonmigrating antioxidants have been shown to be effective at the lower temperatures usually encountered during long-term aging. Propagating radicals could reach inhibition sites by movement along polymer chains as suggested in Sect. C-3, or as a result of chain flexibility. Reactivity of radicals with polymer molecules is also reduced at these temperatures thus providing more opportunity for stabilization reactions to take place. It is quite probable that a combination of these factors is responsible for the inhibition of oxidation by nonmigrating antioxidants.

a) Carbon Black as a Thermal Antioxidant

Carbon black is perhaps the most versatile additive used to inhibit polymer degradation. It is most frequently employed as a stabilizer to inhibit photooxidation, but as was suggested previously, it also functions as a thermal antioxidant. As a polymer additive, carbon black is used in the form of bulky particles in contrast to the molecular dispersion of other antioxidants. The average particle size is in the 15 to 25 mμ range, and the concentration is normally between 2.5 and 3.0 percent[59] when used as a photostabilizer.

The structure of carbon black, shown schematically in Fig. D-9, suggests several ways in which it could function as a thermal antioxidant. The structure of this unique

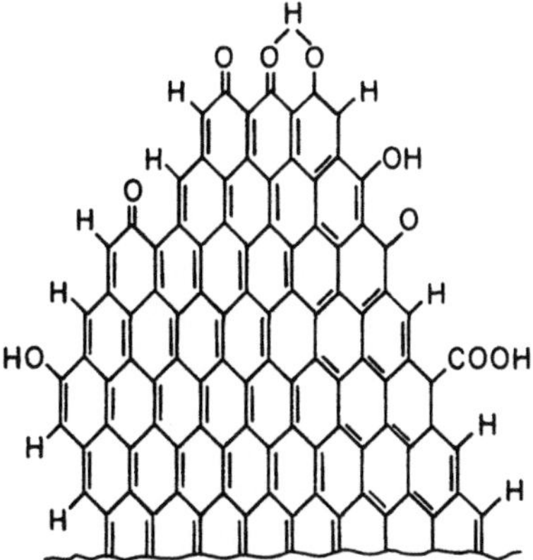

Fig. D-9. Typical structures believed to be present in carbon black. (Reprinted with permission of Harper and Row Publishers, Inc.)

stabilizer consists of conglomerates of many condensed polycyclic aromatic rings. A variety of oxygenated groups are chemically combined on the particle surface, and there are unpaired electrons within the structure. Trapping of propagating radicals could result from the radical nature of carbon black, and it has been shown[60] that it does have the capability for trapping small radicals such as methyl radicals. These radicals are small enough to penetrate into the carbon black structure where most of the unpaired electrons exist. The larger polymer radicals, however, could only be trapped by electrons at the particle surface. Though radical trapping may play a minor role, there is no clear relationship between the number of free electrons, measured by electron spin resonance, and the ability of carbon black to inhibit thermal oxidation. The relationship between the concentration of oxygen-containing, functional groups on the carbon surface and antioxidant activity, on the other hand, has been demonstrated by Hawkins and coworkers[55–57].

Commercial channel blacks have between 3 and 4 percent oxygen combined on the surface, and much of this is believed to be present as phenolic groups. So carbon black can be regarded as a giant phenol and as such should be capable of functioning as a labile-hydrogen donor. As shown in Fig. D-10, commercial channel black inhibits the oxidation of low-density polyethylene at 140 °C for 20–30 h. Heating of channel black at about 900 °C under nitrogen pyrolyzes off almost all combined oxygen, and this modified carbon black no longer functions as a thermal antioxidant. Conversely, heating in an oxygen environment increases the concentration of combined oxygen to 18 percent. This oxygen-activated carbon black is an excellent antioxidant for branched polyethylene at 140 °C and in 3 percent concentration. This activated carbon black has a higher concentration of oxygenated groups on the surface.

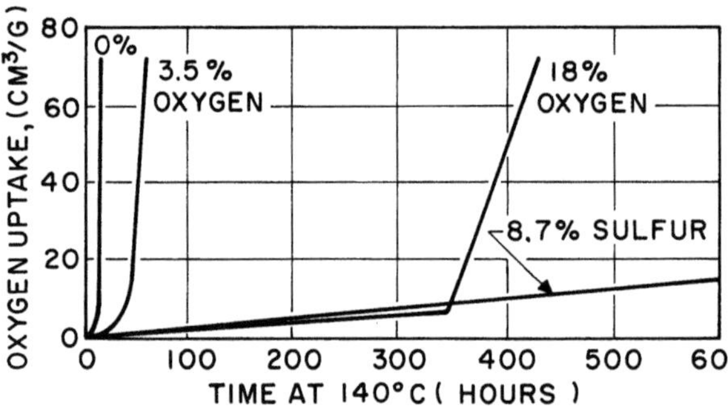

Fig. D-10. Stabilization of polyethylene with activated carbon black. (Reprinted with permission of Wiley-Interscience)

Channel black can also be activated by heating with elemental sulfur. About 8 percent sulfur is chemically bound since it cannot be extracted. Sulfur-activated carbon black is superior even to the oxygen-activated black as seen in Fig. D-10. It retards oxidation for hundreds of hours at 140 °C, with complete suppression of autocatalysis. As will be discussed subsequently, carbon blacks also function as hydroperoxide decomposers with the activated blacks being the most effective. Mercaptan groups are probably present on the surface of sulfur-activated carbon blacks.

A unique temperature effect is observed [55] with branched polyethylene containing 3 percent carbon black as shown in Fig. D-11. Above 100 °C the polymer crystallinity is completely destroyed so that the entire mass is vulnerable to oxidation. If the data taken at elevated temperatures are extrapolated linearly, this formulation would be expected to fail in about 3 years at about 30 °C. However, similar formulations have been shown to resist outdoor aging for over 30 years at average temperatures close to 30 °C. Obviously the linear extrapolation, which serves to predict long-term life for most antioxidants, does not apply to carbon black/polyethylene formulations. The extrapolation deviates from linearity in the region with which branched polyethylene melts. The inserts in this figure offer a possible explanation for these data.

In the molten state, carbon black must protect the entire polymer mass, but below 100 °C about 60 percent of the polymer is protected from rapid oxidation by virtue of its crystalline structure [64]. Since oxygen cannot penetrate the crystallites of polyethylene, oxidation can only take place at their surface and this degradation is quite slow as compared to oxidation in the amorphous regions. Since only 40 percent of the polymer needs to be protected below the T_m, carbon black now becomes very effective. This effect is essentially equivalent to increasing the concentration of carbon black in the formulation. The effect of increased concentrations of carbon in inhibiting polyethylene oxidation is shown in Fig. D-12. Carbon blacks, made by the furnace process have less

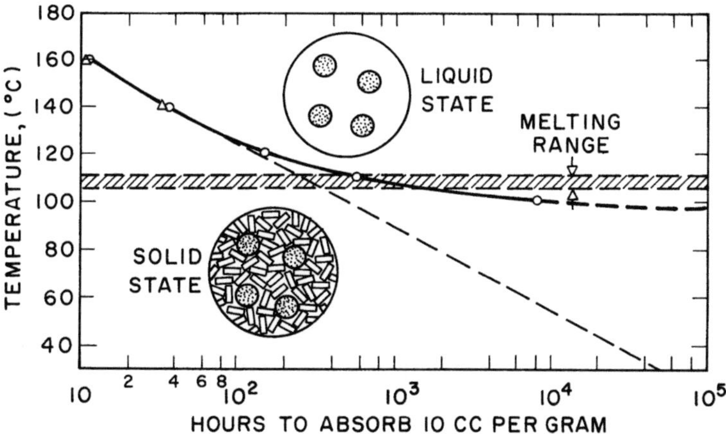

Fig. D-11. The temperature effect in the oxidation of low-density polyethylene stabilized with 3% carbon black.(Reprinted with permission of The American Chemical Society)

functional groups on the surface but show proportionally similar effects to those exhibited by channel blacks [65].

Hawkins and Winslow [55] synthesized a series of complexes by coupling polyfunctional phenols onto the surface of sulica particles. The particle size of the silica was selected to be about the same as that for particles of carbon black. Coupling to silica involved phenolic groups but a considerable number of these remained free on the surface of the complexes and thus were available for chain-breaking reactions. These complexes functioned as antioxidants as expected, and the temperature effects were parallel to those found for carbon black. Because of their bulky nature, the complexes are nonmigratory, and, like carbon black, become very effective as polymer morphology changes during crystallization. However, the complexes decomposed slowly on standing and have not been used commercially.

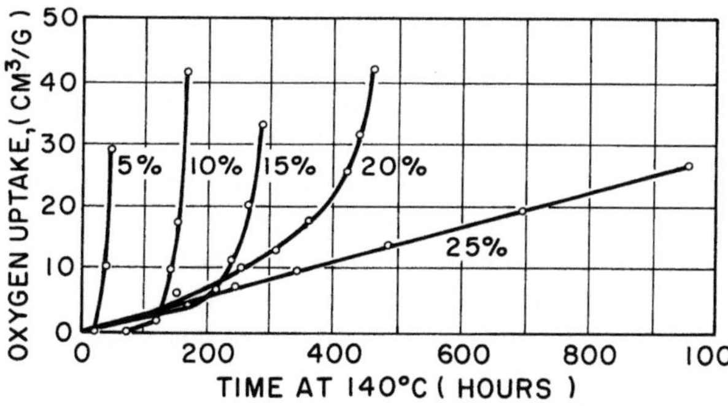

Fig. D-12. Thermal oxidation rate of polyethylene containing various amounts of carbon black. (Reprinted with permission of John Wiley and Sons, Inc.)

Carbon black also effectively catalyzes hydroperoxide decomposition into nonradical products. This reaction can be attributed to the acidity resulting from oxygen-containing groups. It is evident from Fig. D-13 that its activity in hydroperoxide decomposition increases with the amount of chemically-bound oxygen. Acidity also increases as more oxygenated groups are formed on the particle surface. Since sulfurized carbon

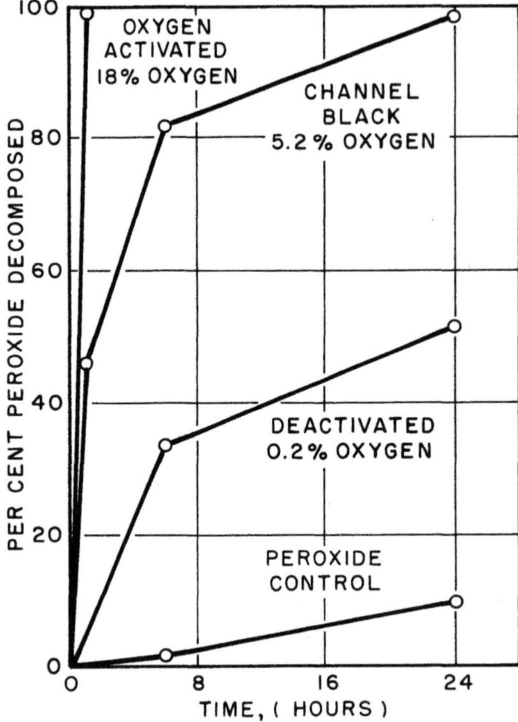

Fig. D-13. Decomposition of cumyl hydroperoxide in the presence of various carbon blacks

black is even more effective than the oxygen-activated carbon blacks, it is reasonable to assume that sulfurized groups undergo reactions similar to those described for sulfur-containing, preventative antioxidants.

In summary, carbon black can function as a thermal antioxidant by several different mechanisms. It probably reacts to a minor extent by trapping propagating radicals. Its primary reaction in chain-breaking is as a labile-hydrogen donor, particularly with the activated carbon blacks. Because of the unusual temperature effects resulting from the bulky size of carbon black particles, it is very effective as an antioxidant in solid, semi-crystalline hydrocarbon polymers. Granted that concentrations are much higher than those used for molecular-dispersed antioxidants. Carbon black must be used in 2.5 to 3 percent concentration to effectively protect against photodegradation.

Carbon black is used in from 2.5 to 3 percent concentration for protection against ultraviolet-induced radiation. Although this is the primary function of carbon black in stabilization, its role as a chain-terminating and preventative antioxidant cannot be neglected. Finally, the ability of carbon black to react synergistically with other antioxidants is important in many applications. The only restriction to use of this versatile additive is that black formulations must be acceptable.

b) Bound or Grafted Antioxidants

The chemical bonding of antioxidants onto polymer molecules would prevent migration and the accompanying loss of stabilizers. A number of reactions have been proposed and several tried in attempts to graft antioxidants to polymer chains. Antioxidants bound in this way would not be lost either through evaporation or solvent extraction. Positive results with such nonmigrating antioxidants were first reported in studies on the stabilization of natural rubber.

N,N'-diethyl-p-nitrosoaniline (XIII) inhibits the oxidation of squalene, a model for natural rubber, but only after a finite induction period [66]. Preheating of squalene with the nitroso compound in the absence of oxygen, however, results in immediate stabilization. The following reactions were suggested to account for these observations,

$$-\left(-CH_2-\overset{\overset{\displaystyle CH_3}{|}}{C}=CH-CH_2-\right)_x- \; + \; (C_6H_5)_2NC_6H_4N=O \longrightarrow -\left(-CH=\overset{\overset{\displaystyle CH_3}{|}}{C}-\underset{\underset{\displaystyle NHC_6H_4N(C_6H_5)_2}{|}}{C}H-CH-\right)_x-$$

$$\text{XIII}$$

The addition product from this reaction is a p-phenylenediamine, and secondary amines of this type are known to be excellent antioxidants for hydrocarbon polymers. The diamine moiety could not be solvent extracted, confirming that it was bonded to the squalene. The significance of bonding the antioxidant is evident in Table D-4. Both nitrosophenols and nitrosoamines can be bonded to rubber by this reaction. In each case, extensive aqueous extraction did not remove the antioxidant. Under similar extraction conditions and with conventional compounding, the p-phenylene diamine would have been almost totally lost. The conventional, short-term antioxidant, 2,6-ditert.-butyl-p-cresol, lost almost all of its effectiveness on extraction, indicating that it was not bonded to the polymer. Antioxidants bound through nitroso groups inhibit as effectively at low temperatures as the corresponding p-phenylenediamines or substituted phenols when added to the polymer by conventional techniques.

Table D-4. Additives incorporated into natural rubber and as bound antioxidants

Additive	Hours to absorb 1% by wt. of O_2	
	before extraction	after extraction
N,N'Diethyl-p-nitrosoaniline	39	30
p-Nitrosodiphenylaniline	60	53
p-Nitrosophenol	31	30
2,6-Ditert.-butyl-p-cresol (Not bonded)	47	4

Kaplan and coworkers [67, 68] have bonded various phenols to polyethylene and to polyoxymethylene by the following sequence of reactions,

OH / NO$_2$ $\xrightarrow{\text{Reduction over Sn}}$ OH / NH$_2$ $\xrightarrow{\text{Diazotization HONO}}$ O / N$_2$ (XIV)

O / N$_2$ $\xrightarrow{\text{Heat}}$ O (XV)

Bonding of the stabilizer moiety to the polymer could occur in several ways. One possibility is that the diazooxide (XIV) could decompose thermally into a carbene (XV) which could then be inserted into C—H bonds in the polymers by known reactions [69],

Alternatively, insertion of the antioxidant moiety could have taken place by a one-step, singlet reaction or the diazooxide (XIV) could undergo an abstraction-recombination reaction bonding to the polymer without going through the carbene step [70].

Each of the samples, treated with XIV was extracted with boiling water for 24 h before being subjected to reaction with oxygen. After this extraction, high-density polyethylene was more stable to oxidation than commercially-stabilized polymer which lost much of its stability during a comparable extraction. Similar results were obtained when polyoxymethylene was treated with XIV. Although the exact mechanism has not been firmly established, the efficiency of bonded antioxidants is apparent from Table D-5. Reaction was carried out by adding 3,5-di-tert. butyl-1,4-diazooxide to the polymers in solution. The solvent was then removed and test samples were molded. Apparently bonding took place during the molding process.

Table D-5. Stability of hydrocarbon polymers with bound phenolic antioxidants

	Hours to react with 10 cc. of O_2 at 140 °C
Low-density Polyethylene:	
Uninhibited	3
Reacted with XIV	14
High-density Polyethylene:	
Commercially stabilized	175
Reacted with XIV	411
Polypropylene:	
Uninhibited	< 1
Reacted with XIV	31

XIV is 3,5-Ditert.butyl-1,4-diazooxide

Scott and coworkers [71-73] have succeeded in grafting sulfur-based antioxidants to acrylonitrile-butadiene-styrene (ABS) resins. Yields as high as 50 percent of grafted ABS were obtained. High temperature and a high concentration of the antioxidant during pretreatment are required for high yields of the grafted polymer. These macromolecular antioxidants have been used as masterbatches to be added to unmodified ABS. Sulfur-based antioxidants, as represented by XVI, were grafted to the polymer by addition to double bonds, either using free-radical precursors or through radicals formed during processing [72],

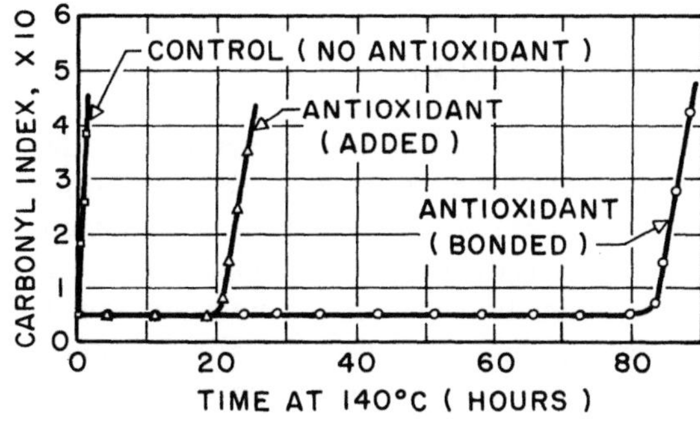

Oxidation by hydroperoxides then yields acidic sulfur products which function as effective preventative antioxidants. Disulfides may also form from XVI during the reaction, and these could decompose thermally into sulfenyl and sulfinyl radicals,

$$2\,RSH \rightarrow RSSR \rightarrow R\overset{\overset{\textstyle O}{\|}}{S}SR \rightarrow RS^{\bullet} + RSO^{\bullet}.$$

Either of these radicals could then become bound to the polymer and function as hydroperoxide decomposers.

Adduct yields of 68 percent were obtained when XVI mixtures with polypropylene were subjected to uv radiation[71]. As in the grafting of antioxidants to ABS, reaction also takes place during processing. The bound antioxidant effectively inhibits the thermal oxidation of polypropylene, approaching the level of protection provided by blending the most effective commercial antioxidants into the polymer. After solvent extraction, the bound antioxidant was far superior as shown in Fig. D-14. Grafting of antioxidants to polypropylene has considerable commercial importance since this very sensitive polymer oxidizes rapidly once the antioxidant is depleted.

Fig. D-14. Comparison of thermal-oxidative stability at 140 °C of polypropylene films containing 3,5-DI-T-butyl-4-hydroxybenzyl mercaptan as a bound antioxidant and as an additive at the same concentration, ca. 0–15%. (Reprinted with permission of Applied Science Publishers, Ltd.)

In theory at least, it should be possible to incorporate antioxidants into polymers by copolymerization, using a comonomer with antioxidant capability, as for example, XVII,

D) Stabilization Against Thermal Oxidation

However, polymerization is likely to be retarded if propagating radicals are deactivated by labile hydrogens in the comonomer. These hydrogens could be masked temporarily by esterification, but then hydrolysis would be required once polymerization was completed. For this reason, grafting rather than copolymerization is favored as a means of incorporating antioxidants into a polymer.

Photostabilizers have been bound to polymer molecules by Vogel and his associates [74]. Polymerizable photostabilizers have also been prepared as represented by XVIII and XIX, in which the ethylenic double bonds provide sites for polymerization.

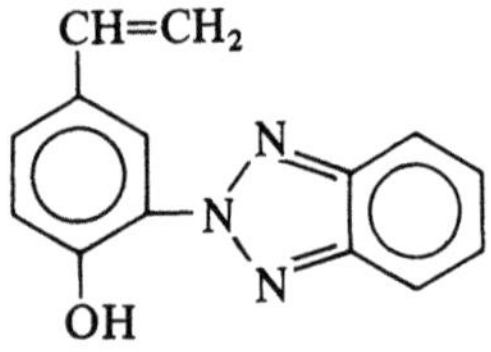

CH=CH$_2$ CH$_3$—C=CH$_2$

XVIII 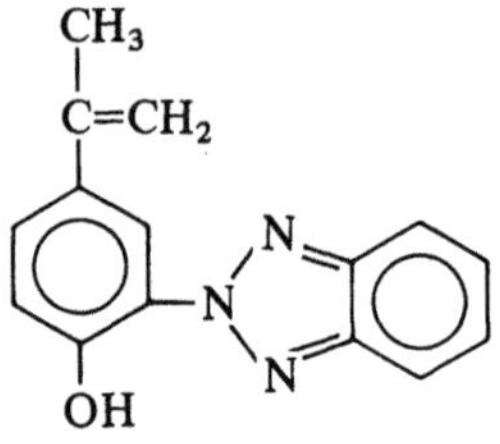   XIX

Compound XVIII has been homopolymerized and can be copolymerized with styrene and acrylic monomers. In contrast, XIX forms copolymers but it has not as yet been homopolymerized. This emphasizes some of the difficulties encountered in polymerizing monomers containing an antioxidant moiety.

Whether a nonmigrating antioxidant is joined to the polymer by grafting or copolymerization, its immobility limits, its effectiveness in stabilization at high temperatures. Short-term antioxidants would still be needed to protect very sensitive polymers at the higher temperatures used for processing. Furthermore, any modification of a polymer's structure, as in the bonding of stabilizers, may alter other properties beyond acceptable limits.

6) Stabilization by Structure Modification

Polymers that are stable to thermal degradation because of their structure were discussed in Sect. C-I. Stability to oxidative degradation might be attained by following the general principles used to develop heat resistant polymers. However, such obvious approaches as incorporating bonds of higher dissociation energy, developing structures that would be more highly crystalline, or avoiding branching in the molecules usually results in the loss of other important properties. This section will deal with the limited methods which would preserve all essential properties of a polymer while making minor changes in its structure to improve stability.

Oxidative degradation is limited by the rate of oxygen diffusion into a polymer. If sufficient oxygen cannot migrate below the surface layers to maintain the maximum oxidation rate, reaction will become diffusion controlled and degradation will be limited to surface layers. It follows then that structural changes that tend to reduce diffusion of oxygen into the polymer bulk offer a route to improving stability. Two methods have been considered to reduce oxygen penetration into polymers.

In many semicrystalline polymers, e.g. polyethylene and polypropylene, crystalline regions are so densely packed that penetration of oxygen into the polymer is restricted.

However, sufficient oxygen migrates through amorphous regions to cause significant degradation. Crystallization of molten polyethylene against noble metal surfaces [76] has been investigated as a method for generating a dense, transcrystalline layer at the surface. The denser crystalline structure would be expected to restrict oxygen penetration into the polymer bulk. This technique, however, was not successful since the transcrystalline surface layer rapidly reverted to the normal crystalline structure on standing at room temperature.

Cross-linking is a more promising approach to stabilization by structure modification. A dense, cross-linked surface layer would restrict oxygen diffusion into the polymer bulk. Radiation or reaction with peroxides could be used to cross-link just at the surface. For some applications, however, even this surface modification may result in unacceptable property alteration. Though improving a polymer's stability to oxidative degradation by surface cross-linking is theoretically promising, it has not found significant commercial application.

7) Stabilization Against Burning

The burning of polymers is a special case of thermal oxidation. Both structure modification and additives have been used to improve the resistance of polymers to burning. Induced char formation at the polymer surface under burning conditions has been investigated [77, 78] as a technique for improving resistance to burning. These chars are reasonably stable barriers which reduce the escape of combustible volatiles into the burning zone. Char formation takes place through condensation reactions in which a variety of catalysts have been used [77, 78] as promoters. Intumescent coatings are a special class of foamed or expanded chars. As with many techniques for improving resistance to burning, additives in large amounts often alter properties of the polymer beyond acceptable limits.

Antimony oxide is one of the most effective additives for improving resistance to burning. This additive is only effective when chlorine is present. Reaction with chlorine generates antimony trichloride which then escapes into the burning zone. Other additives, including compounds of phosphorous and halogens, bromine in particular, also function after they or their decomposition products are volatilized into the burning zone. In the condensed phase, flame retardants function by absorbing heat in the polymer bulk through endothermic reactions. The absorption of heat in the dehydration of inorganic hydrates such as alumina hydrate is a typical example of the physical inhibition of combustion in the condensed phase. Inorganic fillers also function as noncombustible diluents for combustible polymers. Stabilization of polymers under combustion conditions is a broad subject on which numerous reviews have been written [77-80].

II) Nonhydrocarbon Polymers

Mechanisms for stabilization of nonhydrocarbon polymers have not been developed to the same extent as those for the hydrocarbon polymers. In those instances where a polymer contains a significant hydrocarbon segment, oxidation by the radical-initiated, chain mechanism can contribute to the overall degradation mechanism. In other non-hydrocarbon polymers, different reaction mechanisms are primarily responsible for degradation.

Polymers that are formed by condensation reactions are susceptible to degradation by hydrolysis. Reaction with water cleaves primary bonds between repeating units in the polymer chain, reducing the molecular weight. Stabilization of these polymers against hydrolysis would require protection against acidic or basic catalysts that promote hydrolysis. Stabilizers which can neutralize these catalysts are effective. Acidic products formed by thermal oxidation can also catalyze hydrolysis[81]. When such products are formed, it is very difficult to adjust the concentration of basic stabilizers for optimum protection. An excess of the base could promote hydrolysis before reacting with the acidic products that are being formed.

Hydrolysis of pendant groups on the polymer chain, e.g. ester groups in poly(methyl methacrylate), can also lead to degradation,

$$
\left[\begin{array}{c} \overset{O}{\overset{\|}{C}}-OCH_3 \\ | \\ -C-CH_2- \\ | \\ H \end{array} \right]_n \ \xrightarrow{\ HOH\ } \ \left[\begin{array}{c} \overset{O}{\overset{\|}{C}}-OH \\ | \\ -C-CH_2- \\ | \\ H \end{array} \right]_n \ + \ n \ CH_3OH
$$

Although these reactions do not result in chain scission they do alter molecular structure, and significant changes in properties would occur at high levels of hydrolysis.

Nylon 11 (XX) has a chain of ten methylene groups in each repeating unit,

$$
-NH-\left[-(CH_2)_{10}-\overset{O}{\overset{\|}{C}}- \right]_n - \qquad XX
$$

and oxidative degradation can occur at these hydrocarbon segments. As might be expected, conventional antioxidants are effective stabilizers. Dependent on exposure conditions, either hydrolysis or oxidation could be the predominant mechanism of degradation in this polymer. In all probability, both reactions would take place simultaneously in Nylon 11. Other nylons with shorter hydrocarbon segments are less vulnerable to oxidation and would probably degrade only by hydrolysis. The extent to which conventional antioxidants can protect a nonhydrocarbon polymer is related to the number and length of hydrocarbon segments in the structure.

The stabilization of poly(vinyl chloride) (PVC) is an interesting example in which complex degradation reactions, including oxidation, are involved. The principal mechanism of degradation for this polymer is a sequential elimination of hydrogen chloride (HCl) (refer to B-I-2-a). However, several different oxidation reactions are important in the overall degradation mechanism. As HCl is split off, a polyene product is formed. This polyene could undergo typical oxidation reactions, leading to chain scission and/or cross-linking. Mechanical properties are affected as these reactions proceed. Conventional antioxidants will inhibit these secondary reactions and minimize the loss in mechanical properties. However, in many PVC applications, discoloration resulting from HCl elimination causes failure before there is any significant change in mechanical properties.

68

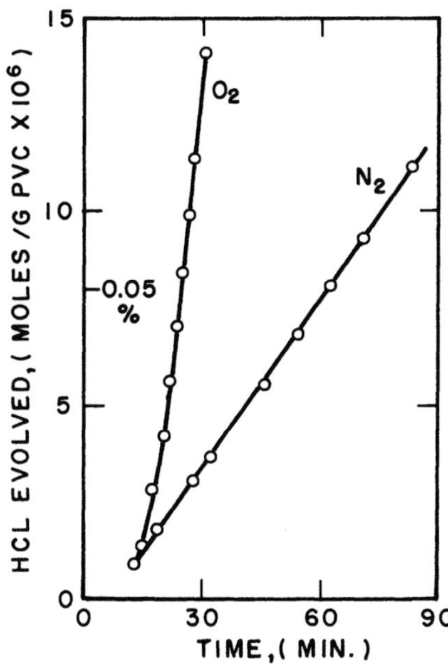

Fig. D-15. Effect of O_2 on HCl elimination. (Reprinted with permission of Wiley-Interscience)

Oxygen catalyzes dehydrochlorination as shown in Fig. D-15. The following reactions have been suggested to account for the role of oxygen in PVC degradation,

$$- - -CH = CH\overset{\bullet}{C}CHCl- - - \xrightarrow{O_2} - - -CH = CHCHCl- - -$$
$$|$$
$$OO^\bullet$$

$$\xrightarrow{PVC} - - -CH = CHCHCHCl- - - + - - -\overset{\bullet}{C}H-CHCl- - -$$
$$|$$
$$OOH$$

$$- - -CH = \overset{\downarrow}{C}HCHCHCl- - - \xrightarrow{PVC}$$
$$|$$
$$O^\bullet$$

$$\rightarrow - - -CH = CHCHCl- - - + PVC .$$
$$|$$
$$OH$$

It has been suggested [35] that oxygen-containing groups, occurring as impurities in PVC, may also function as sites for initiation of dehydrochlorination. Although there is considerable evidence for the role of oxidation in PVC degradation, conventional, chain-breaking antioxidants have very little effect as stabilizers – at least in rigid PVC. Preventative antioxidants which suppress radical formation from hydroperoxide decomposition, however, reduce the rate of HCl elimination as shown in Fig. D-16. Naphthyl disulfide, an effective hydroperoxide decomposer, reduces the rate of HCl elimination, but the chain-breaking antioxidants dinaphthyl-p-phenylenediamine and 4,4'-methylenebis(2-tert. butyl-5-methylphenol) have very little effect.

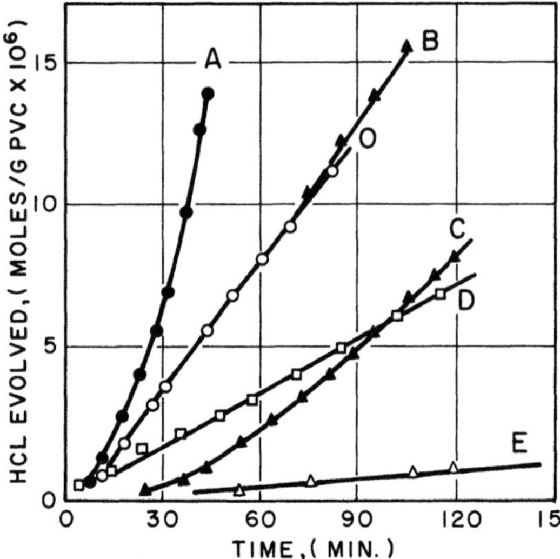

Fig. D-16. Effects of various stabilizers on dehydrochlorination of PVC under nitrogen: A, DI-β-naphthyl-p-phenylene diamine; B, 2,2'-methylene bis(4-methyl-6-t-butylene phenol); C, naphthyl disulfide; D, dibutyl tin octoate; E, barium stearate. (Reprinted with permission of Wiley-Interscience)

In contrast to rigid PVC, thermal oxidation can be very important in the degradation of flexible (plasticized) PVC. Most commercial plasticizers, as represented by dioctyl phthalate (XXI), have large hydrocarbon segments.

$$\text{XXI}$$

Oxidation of these plasticizers proceeds by typical hydrocarbon reactions with eventual loss of plasticizer efficiency. Esters of azelaic and sebacic acids may undergo oxidative reactions in both the acid and alcohol segments. Chain-breaking antioxidants, e.g. Bisphenol A, are effective as inhibitors of plasticizer oxidation, and are important in stabilizing highly-plasticized PVC formulations. In addition to the primary plasticizers, polymeric additives are sometimes used as secondary plasticizers in PVC. Partially chlorinated polyethylene is a typical example of this class of plasticizers. Since these plasticizers are largely hydrocarbon in composition, conventional antioxidants are important in preventing their oxidative degradation.

Although oxidative reactions are important in PVC degradation, especially in flexible formulations, the principal reaction responsible for degradation occurs through HCl elimination. Therefore additives which offset the catalytic action of liberated HCl are the most important stabilizers for PVC. There have been several recent reviews [82–84] on the stabilization of PVC using acid absorbers.

Many basic compounds have been used as stabilizers for PVC, lead compounds being the first to be applied commercially. Despite their early applications and low cost, lead compounds are not widely used today because of their toxicity. Also lead compounds reduce optical clarity and so are not used in clear or transparent formulations. Basic lead carbonate (white lead), tribasic lead sulfate, lead stearate and other lead compounds have gadually been replaced by salts or organometallic complexes of barium, cadmium and tin. Although these stabilizers are more expensive, they are not as toxic as lead compounds. They also have other important advantages. For example, dibutyl tin dilaureate has good uv stability and is therefore suitable for clear formulations to be used out-of-doors. Tin mercaptides have good heat and uv stability and are widely used in transparent compounds. Some of the newer acid acceptors also function to a minor extent as hydroperoxide decomposers.

Oxidation of partially dehydrochlorinated PVC can reverse low levels of discoloration. Conjugated sequences that produce color are disrupted as the polyene structure is oxidized. This accounts for the "bleaching" of slightly-degraded PVC. Although this reaction retards color formation, it could lead to loss in mechanical properties.

In summary, complete stabilization of PVC must take into account all reactions that cause degradation of this polymer. Acid absorbers reduce the catalytic effect of liberated HCl and antioxidants suppress HCl elimination as well as inhibiting plasticizer oxidation in flexible formulations.

References

1. Bateman, L.: Quart. Rev., *8*, 147 (1954)
2. Bolland, J.L.: Quart. Rev., *3*, 1 (1949)
3. Cunneen, J.I.: Rubber Rev., Rubber Chem. Technol., *41*, 182 (1968)
4. Semon, W.L.: History and Use of Materials which Improve Aging, in: The Chemistry and Technology of Rubber (eds) Davis, C.C., Blake, J.T., p. 414, New York: Reinhold 1937
5. Ostwald, W.: J. Soc. Chem. Ind. (London), *32*, 179 (1913)
6. Moureau, C., Dufraisse, C.: Bull. Soc. Chim. (France), *31*, 1152 (1922)
7. Moureau, C., Dufraisse, C.: Chem. Rev., *3*, 113 (1926)
8. Fisher, H.L.: Chem. Rev., *7*, 130 (1930)
9. Bodenstein, M.: Z. Phys. Chem., *84*, 329 (1913)
10. Taylor, H.S.: J. Phys. Chem., *27*, 322 (1923)
11. Christiansen, J.A.: J. Phys. Chem., *28*, 145 (1924)
12. Bäckström, J.A.: J. Am. Chem. Soc., *49*, 1460 (1927)
13. Chakraborty, K.B., Scott, G.: Polym. Degrad. and Stab., *1 No 1*, 37 (1979)
14. Ambelang, J.C. et al.: Rubber Rev., Rubber Chem. and Technol., *36*, 1497 (1963)
15. Barnard, D. et al.: Oxidation of Olefins and Sulfides, in: Chemistry and Physics of Rubberlike Substances (ed) Bateman, L., p. 593, Macheren, London 1963
16. Price, C.C.: J. Chem. Ed., *42(1)*, 13 (1965)
17. Bowden, M.J.: Formation of Macromolecules, in: Macromolecules (eds) Bovey, F.A., Winslow, F.H., p. 28, Academic Press, London 1979
18. Ingold, K.U.: Inhibition of Oxidation, Advances in Chemistry Series, Amer. Chem. Soc., *77*, 296 (1968)
19. Barton, D.H.R., Howlett, J.: J. Chem. Soc., 144 (1960)
20. Boozer, C.E., Hammond, G.S.: Amer. Chem. Soc., *76*, 3861 (1954)
21. Hammond, G.S., Boozer, G.E., Hamilton, C.E.: J. Amer. Chem. Soc., *77*, 3238 (1955)
22. Ingold, K.U., Puddington, J.E.: Ind. Eng. Chem., *51*, 1319 (1959)
23. Thomas, J.R., Tolman, C.A.: J. Amer. Chem. Soc., *84*, 2930 (1962)
24. Shelton, J.R., McDonel, E.T.: J. Polym. Sci., *32*, 75 (1960)
25. Shelton, J.R., Vincent, D.N.: J. Amer. Chem. Soc., *85*, 2433 (1963)
26. Ingold, K.U., Howard, J.A.: Nature, *195*, 280 (1962)
27. Reich, L., Stivala, S.S.: Autooxidation of Hydrocarbons and Polyolefins, New York: Dekker 1969

D) Stabilization Against Thermal Oxidation

28. Loan, L.D., Winslow, F.H.: Reactions of Macromolecules, in: Macromolecules (eds) Bovey, F.A., Winslow, F.H., p. 422, New York: Academic Press 1979
29. Bickel, A.F., Kooyman, E.C.: J. Chem. Soc., 3211 (1953)
30. Dunn, J.R., Scanlan, J.: Trans. Inst. Rubber Ind. (London), *34*, 228 (1958)
31. Dunn, J.R., Scanlan, J.: J. Polym. Sci., *35*, 267 (1959)
32. Kinnerly, G.W., Patterson, W.L., Jr.: Ind. Eng. Chem., *48*, 1917 (1956)
33. Kharasch, M.S., Nudenberg, W., Mante, G.J.: J. Org. Chem., *16*, 524 (1951)
34. Shelton, J.R.: Thermal Oxidation of Polymers, in: Stabilization and Degradation of Polymers, Advances in Chemistry Series, Amer. Chem. Soc., *169*, 218 (1978)
35. Scott, G.: Pure Appl. Chem., *30*, 267 (1972)
36. Scott, G.: Eur. Polym. J., *11*, 161 (1975)
37. Hawkins, W.L., Sautter, H.: Chem. Ind. (London), 1825 (1962)
38. Hawkins, W.L., Sautter, H.: J. Polym. Sci., Part 1A, 3499 (1963)
39. Chasar, D.W.: Polym. Degrad. and Stab., *3 No. 2*, 121 (1981)
40. Colcough, T., Cunneen, J.I.: J. Chem. Soc., 4790 (1964)
41. Burn, A.J.: Tetrahedron, *22*, 2153 (1966)
42. Lucken, E.A.C.: J. Chem. Soc., 1354 (1966)
43. Burn, A.J.: Mechanisms of Oxidation Inhibition by Zinc Dialkyl Dithiophosphates, in: Oxidation of Organic Compounds, Advances in Chemistry Series, Amer. Chem. Soc., *75(I)*, 323 (1968)
44. Ivanov, S.K.: Mechanism of the Dithiophosphate Antioxidants, in: Polymer Stabilization-3 (ed) Scott, G., p. 55, Applied Science Publishers, Ltd., England 1980
45. Hansen, R.H., DeBenedictis, T., Martin, W.H.: Trans. Inst. Rubber Ind. (London), *39*, 290 (1963)
46. Chan, M.G., Allara, D.L.: Polym. Eng. Sci., *14*, 12 (1974)
47. Gould, E.S., Rado, M.: L. Catalysis, *13*, 238 (1969)
48. Hansen, R.H. et al.: J. Polym. Sci., *Part 2A*, 587 (1964)
49. Calvin, M., Bailes, R.H.: J. Amer. Chem. Soc., *68*, 953 (1946)
50. Allara, D.L., Chan, M.G.: J. Polym. Sci., *14*, 1857 (1976)
51. Allara, D.L., White, C.W.: Microscopic Mechanisms of Oxidative Degradation and its Inhibition at a Copper-Polyethylene Interface, in: Stabilization and Degradation of Polymers, Advances in Chemistry Series, Amer. Chem. Soc., *169*, 273 (1978)
52. Hawkins, W.L. et al.: J. Appl. Polym. Sci., *1 No. 1*, 37 (1959)
53. Scott, G.: Atmospheric Oxidation and Antioxidants, p. 204, Amsterdam: Elsevier 1965
54. Shelton, J.R.: Stabilization Against Thermal Oxidation, in: Polymer Stabilization (ed) Hawkins, W.L., p. 107, New York: Wiley-Interscience 1972
55. Hawkins, W.L., Winslow, F.H.: Trans. Plastics Inst. (London), *29*, 82 (1961)
56. Hawkins, W.L., Worthington, M.A.: J. Polym. Sci., *1 Part A*, 3493 (1963)
57. Hawkins, W.L., Winslow, F.H.: Degradation and Stabilization, in: Crystalline Olefin Polymers (eds) Raff, R.V.A., Doak, K.W., p. 388, New York: Wiley 1964
58. Knapp, G.G., Orloff, H.D.: Amer. Chem. Soc., Polymer Preprints, General Papers, *5(1)*, 11 (1960)
59. Wallder, V.T. et al.: Ind. Eng. Chem., *42*, 2320 (1950)
60. Moynihan, J.T.: Soc. Plastics Engs., *13 No. 2*, 23 (1957)
61. Studebaker, M.L.: Rubber Chem. Techn., *30*, 1438 (1957)
62. Hallum, J.V., Drushel, H.V.: J. Phys. Chem., *62*, 110 (1958)
63. Garten, V.A., Weiss, D.E.: Rev. Pure Appl. Chem. (Australia), *7*, 67 (1957)
64. Winslow, F.H. et al.: Chem. and Ind., 533 (1963)
65. Chan, M.G., Johnson, L.: Paper presented at The Soc. Plastics Eng. ANTEC Meeting, 1980
66. Cain, M.E. et al.: Rubber J., *150(11)*, 10 (1968)
67. Kaplan, M.E. et al.: J. Polym. Sci., *II No. 6*, 357 (1973)
68. Kaplan, M.L., Kelleher, P.G.: US Patent No. 3,723,405, 1973
69. Kirmse, W.: Carbene Chemistry, 2nd Ed., 209 (1971)
70. Kaplan, M.L., Roth, H.D.: Chem. Comm. 970 (1972)
71. Scott, G., Fauzi Yusoff, M.: Polym. Degrad. and Stab., *3 No. 1*, 53 (1980)
72. Ghaemy, M., Scott, G.: Polym. Degrad. and Stab., *3 No. 6*, 405 (1980–81)
73. Scott, G., Tavakoli, S.M.: Polym. Degrad. and Stab., *4 No. 4*, 267 (1982)

74. Bailey, D., Vogel, O.: J. Macromol. Sci., Rev. Macromol. Sci., *C14(2)*, 267 (1976)
75. Geddis, W.C.: Eur. Polym. J., *3*, 733 (1967)
76. Schonhorn, H.: J. Polym. Sci., *B-2*, 465 (1964)
77. Warren, P.C.: Stabilization Against Burning, in: Polymer Stabilization (ed) Hawkins, W.L., p. 346, New York: Wiley-Interscience 1972
78. Lyons, J.W.: The Chemistry and Uses of Fire Retardants, p. 257, New York: Wiley-Interscience 1970
79. Fenimore, C.P., Martin, F.J.: Mod. Plastics, *43*, 141 (1966)
80. Fenimore, C.P., Jones, G.W.: Combust. Flame, *10*, 295 (1966)
81. Kern, W. et al.: Angew. Chem., *73*, 177 (1961)
82. Sarvetnick, H.A.: Poly(vinyl Chloride), New York: Van Nostrand 1969
83. Starnes, W.H., Jr.: Recent Fundamental Developments in the Chemistry of Poly(vinyl chloride) Degradation and Stabilization, Advances in Chemistry Series, Amer. Chem. Soc., *169*, 309 (1978)
84. Loan, D.L., Winslow, F.H.: Thermal Degradation and Stabilization, in: Polymer Stabilization (ed) Hawkins, W.L., p. 117, New York: Wiley-Interscience 1972
85. Loan, L.D.: unpublished results

E) Stabilization Against Degradation by Radiation

Radiation-induced degradation is a general phenomenon, responsible for failure of polymers in many applications. Absorption of ultraviolet (uv) radiation during outdoor exposure rapidly degrades most polymers, unless they are properly stabilized. However, there are exceptions, e.g. poly(methyl methacrylate) and polytetrafluoroethylene, that resist this type of degradation by virtue of their molecular structure.

As was discussed in Sect. B, most polymers are sensitive to uv radiation in the wavelength region between 300 and 360 nm. Individual polymers absorb and are degraded by uv within a much narrower region, often referred to as the activation spectra maximum [1]. These maxima vary with polymer structure and represent the wavelength region at which each polymer is most susceptible to uv degradation.

Artificial sources of radiation also degrade polymers. X-rays and other electromagnetic sources which generate radiation at damaging wavelengths, including nuclear reactors and radioactive isotopes, can cause polymer degradation, but these sources are of less general importance than is outdoor exposure to solar radiation. Extra-terrestrial applications of polymers are increasing, and the intense solar radiation beyond the earth's atmosphere presents a new and important environment in which degradation could take place. Stabilization against degradation resulting from exposure to each of these sources of radiolytic energy is of considerable importance.

There are three general approaches to stabilization against radiolytic degradation: (1) by blocking or screening out the incident radiation, (2) by the use of additives which preferentially absorb damaging radiation and dissapate the energy in a harmless way, and (3) by the use of additives which deactivate reactive species or intermediates in the polymer as it undergoes degradation. Each of these approaches is exemplified in the stabilization of polymers against uv-induced radiation, the major energy component responsible for degradation during outdoor weathering.

I) Stabilization Against Ultraviolet-induced Degradation

The number of outdoor applications for polymers is constantly increasing. In the construction industry, plastic siding, gutters, window frames and glazing are now commonplace. Many external components of modern automobiles are fabricated from plastics. Pigmentation is used in most of these applications and pigments can provide a measure of stabilization against photodegradation. To the extent that polymer degradation does take place, however, changes in color or shade of pigmented compositions may occur, leading ultimately to failure. Pigments can function as light screens to reduce the penetration of radiation into a polymer matrix, but there is considerable variation among these additives in their efficiency as photostabilizers.

In many applications, pigmentation or coatings cannot be used for aesthetic reasons. Additives for clear formulations must not introduce color or opacity into the polymer. Several different types of photostabilizers have been developed for the protection of clear or light-colored compositions. These stabilizers are classified according to the primary mechanism by which they inhibit photodegradation.

There are three general categories of photostabilizers, each functioning by a different chemical mechanism: (1) uv absorbers, (2) radical traps, and (3) quenchers. Conventional antioxidants used to protect against thermal oxidation can play a role in protecting polymers against uv-induced degradation. These are also used in combination with photostabilizers. In several instances, antioxidants play a dual role as thermal and as photostabilizers. Hindered amines are the latest class of photostabilizers to enter the field. These very effective additives are believed to function by several complex mechanisms, both as thermal and as photostabilizers. Interest is also developing in organometallic complexes, and stabilizers of this type are available.

1) Light Screens

By rigid definition, a true light screen is a surface coating which forms an impenetrable shield between the polymer and the radiation source. Stabilization by a true light screen involves only physical effects. Paints that contain dark pigments function in this way, but the level of protection varies considerably with the nature of the pigment. As a general rule, the darker pigments provide more protection.

Carbon black is the most effective of all light screens, and this pigment also functions as a uv absorber. A continuous film of carbon black over its surface would completely shield a polymer from damaging radiation. However, adequate adhesion of pigments or paints to polymer surfaces is not easy to achieve. Use of the proper vehicle or bonding agent has been used with some success in static applications. Under dynamic conditions involving flexing, however, coatings may lose their adhesion to polymers. As cracks develop in the coating, the shield breaks down, uv radiation reaches the polymer surface, and photodegradation takes place. Better results are obtained when the pigment is dispersed throughout the polymer matrix.

When carbon black is properly dispersed in a polymer, uv radiation reaches only the surface layers. There is very little penetration beyond the immediate surface. Carbon black, however, continues to provide protection below the surface through its ability to absorb that fraction of the uv radiation that penetrates beyond the surface. Since the pigment, in acceptable concentrations, does not exist as a continuous shield at the immediate surface, the first few layers of polymer molecules are exposed to photodegradation. Even this level of degradation can be unacceptable for those applications in which surface conductivity must be held to very low levels. Despite this limitation, carbon black is used widely in the protection of many polymers against photodegradation [2] as shown in Table E-1. There is no other photostabilizer that can retard photodegradation as effectively as carbon black. The only factor restricting even more general use of this stabilizer is the color which it introduces into a polymer.

Table E-1. Effect of carbon black on the outdoor weathering of typical polymers [2]

Polymer	Years required to reach visible surface failure
Polyethylene – clear	1–1½
Polyethylene + 1 pph channel black	> 25
Plasticized poly(vinyl chloride) – clear	1–2
Plasticized poly(vinyl chloride) + 10 pph Channel black	> 15
Neoprene – clear	½–1
Neoprene + 40 pph SRF black	> 20

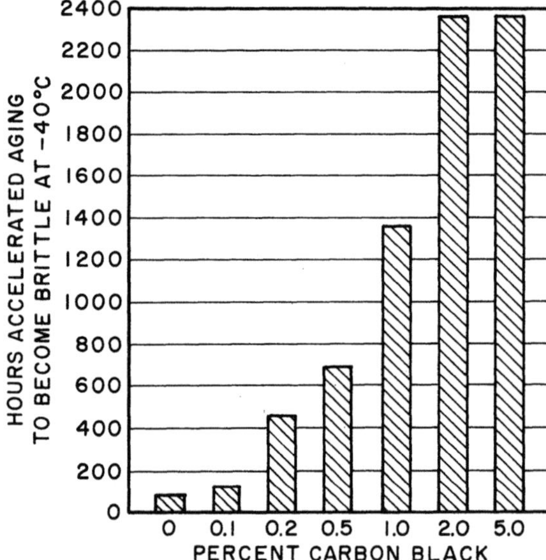

Fig. E-1. Effect of carbon black concentration on accelerated again of polyethylene. (Reprinted with permission of The American Chemical Society)

Wallder and associates [3] have made an extensive study of carbon black in the protection of low-density polyethylene against outdoor weathering. Several important requirements needed for maximum protection were developed in this study. Stability increases with concentration of carbon black up to about 2.0%. Higher concentrations give diminishing levels of protection (Fig. E-1). Furthermore, addition of more than 5% of carbon black can reduce mechanical strength of the formulation beyond acceptable limits. Higher concentrations could be used in a cross-linked polyethylene, but cross-linking also has an adverse effect on certain mechanical properties as for example elongation and flex modulus.

The size of the carbon black particles blended into the polymer is also important. To achieve maximum protection these particles must form a continuous shield within the first few mils below the surface. Larger particle size carbon blacks at equal concentration will not form this continuous layer as near the exposure surface as will smaller particles. The optimum size for carbon black particles was found to be between 15 and 25 mμ. Carbon blacks smaller than this are difficult to disperse uniformly in polyethylene.

The importance of proper dispersion of carbon black in polyethylene is shown in Fig. E-2. At 100 × magnification, a good dispersion appears as a uniform, dark background indicating effective shielding, whereas in the poor (ineffective) dispersion the background is almost white and many large particles or agglomerates of carbon black are evident.

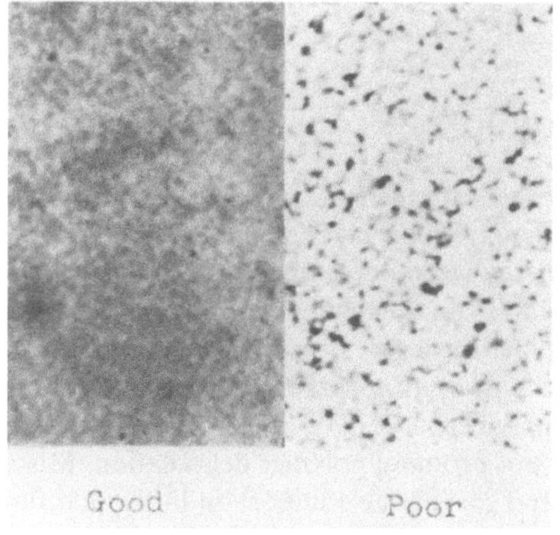

Fig. E-2. Dispersion of carbon black in polyethylene at 100 × magnification. (Reprinted with permission of Wiley-Interscience)

Photodegradation is essentially a surface reaction which is limited by penetration of either uv radiation or diffusion of oxygen into the polymer matrix (refer to Sect. D). A good dispersion of carbon black restricts penetration of uv radiation to only the first few mils below the surface. Oxidation takes place to a much lesser extent than in the clear sample and there is negligible reaction below 10 mils als shown in Fig. E-3. Pro-

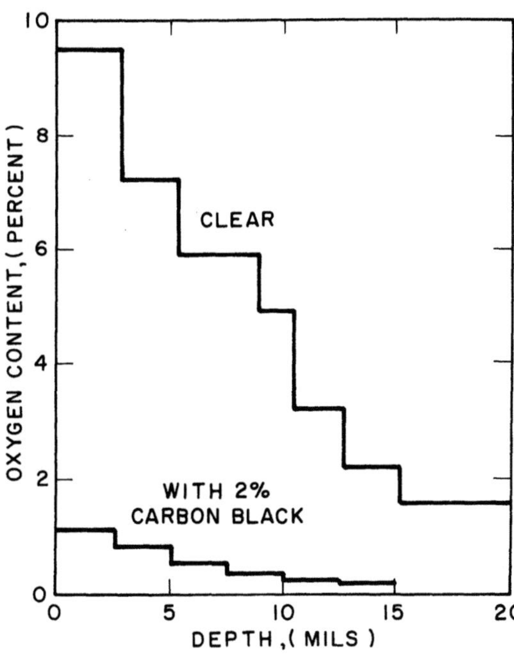

Fig. E-3. Oxygen content of polyethylene at varying depths after 1,000 h under carbon arc. (Reprinted with permission of Elsevier Scientific Publishing Company)

tection at or below this depth can be attributed to absorption of uv radiation by the carbon black. Red iron oxide and other dark pigments have been used to protect polyethylene against photodegradation but these pigments are not as effective as carbon black.

2) Ultraviolet Absorbers

Additives which function as ultraviolet absorbers are the most widely used category of photostabilizers. Although the principal role of carbon black is photostabilization is as a light screen, its ability to absorb uv energy warrants its inclusion in this general category. The most common uv absorbers, however, are low-molecular-weight derivatives of o-hydroxybenzophenone, o-hydroxybenzotriazole or o-hydroxyphenyl salicylate. There are two key requirements for this type of uv absorber: (1) the additive must absorb radiation that would degrade the polymer, and (2) it must dissipate the absorbed energy by a mechanism which does not promote polymer degradation. Miscibility, retention, resistance to degradation and cost are also important in the selection of a uv absorber.

As was discussed in Sect. B, most polymers are sensitive to uv radiation in the wavelength region between 300 and 360 nm. For effective protection, a uv absorber should either absorb radiation over this broad region or have a peak absorption corresponding to the activation spectra maximum of the polymer to be protected. Activation spectra maxima for some typical polymers have been listed in Table B-4. As shown in Fig. E-4, the frequency range of widest sensitivity for polymer degradation is absorbed effectively by the common types of uv absorbers. Selection of a uv stabilizer having a strong peak in its absorption spectrum that closely matches the activation spectra maximum of the polymer to be protected should give the greatest level of protection, all other factors being equal.

Several mechanisms have been proposed through which a uv absorber could dissipate the absorbed energy in a harmless way. Both physical and chemical processes can contribute to energy dissipation. Carbon black, for example, converts absorbed energy into heat by an essentially physical process. Heat that is generated in black formulations contributes to thermal oxidation, however, and antioxidants may be required to provide adequate thermal protection.

Radiative processes provide another physical route for energy dissipation. These processes are based on the conversion of absorbed uv energy into light of longer wavelengths. Fluorescence and phosphorescence provide mechanisms for radiative dissipation of absorbed radiation. However, compounds that fluoresce efficiently are usually decomposed by the absorbed radiation. Therefore the effectiveness of this type of uv absorber is limited by the light stability of the additive. There is also the possibility that light energy emitted as fluorescence or phosphorescence may be reabsorbed by the polymer thus promoting degradation. Despite these limitations, the fluorescing additive, 6,13-dichloro-3,10-diphenyltriphenodioxazine, provides excellent protection for cellulose acetate butyrate[4].

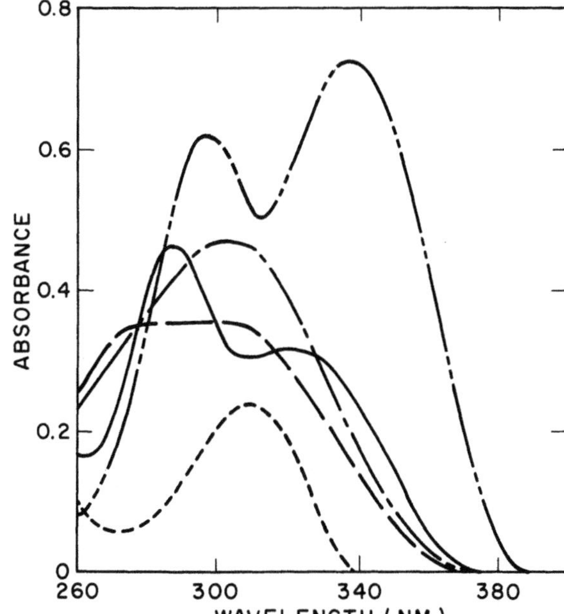

Fig. E-4. Absorbance of uv absorbers at concentration of 1.0 mg/100 ml in chloroform. (Data of P. P. Klemchuk, Ciba-Geigy)

Dissapation of absorbed uv energy can take place by several well established chemical mechanisms. Hydroxybenzophenones (I) are generally believed to function by converting uv energy into vibrational energy within a hydrogen bond. The key to this process lies in the ability of these compounds to form a six-membered ring containing a hydrogen bond. The o-hydroxyphenylbenzotriazoles (II) and o-hydroxyphenyl salicylates (III) probably also dissipate absorbed energy by this mechanism.

The correlation between stabilizer effectiveness and the NMR proton shift in uv absorbers, shown in Table E-2, lends strong support to vibrational energy of the hydrogen bond as the primary mechanism for energy dissipation in these stabilizers. Effectiveness as a photostabilizer increases dramatically with the shift of the hydroxy proton

Table E-2. Correlation between NMR shift and the stabilizing effectiveness of UV absorbers [4]

Additive (1% Concentration)	NMR Hydroxy proton shift	Time required to lose 25% of flexural strength in a modified weatherometer
None	–	200 h
2,6-Dihydroxybenzophenone	– 160	600 h
Phenyl Salicylate	– 220	1,000 h
2,2'-Dihydroxybenzophenone	– 220	1,000 h
2,4-Dihydroxybenzophenone	– 280	2,400 h
2-Hydroxy-4,4'- dimethoxybenzophenone	– 310	2,600 h
3-Benzoyl-2,4- dihydroxybenzophenone	– 340	4,000 h

toward the carbonyl group [4]. Resorcinol monobenzoate (IV) is not in itself an effective photostabilizer, but it rearranges under the influence of uv radiation to 2,4-dihydroxy-benzophenone (V) which can then dissipate harmful radiation by the vibrational mechanism.

IV   V

The effectiveness of uv absorbers is severely reduced or eliminated by any one of the following alterations in molecular structure:

1. Moving the hydroxyl group from the ortho to either the meta or para position – this prevents intramolecular hydrogen-bond formation.
2. Converting the hydroxyl group to a methyl ether – this prevents formation of the hydrogen bond.
3. Insertion of one or more CH_2 groups between the carbonyl group and the phenyl ring containing the orthohydroxyl group – this also hinders intramolecular hydrogen-bond formation since the preferred six-membered ring could not form.
4. Replacement of aromatic rings with cyclohexyl rings – this would diminish the ease of formation of a hydrogen bond.
5. Placement of electron-attracting substituents on the aromatic ring containing the hydroxyl group – this would tend to shift the hydroxyl proton away from the carbonyl group and thus reduce vibration at the hydrogen bond.

A keto-enol tautomerism has been proposed [5] as an extension of the mechanism involving vibration at a hydrogen bond. In this mechanism, heat is released in the reverse step as the tautomerer (VII) reverts back to the original additive (VI).

This reversible rearrangement may supplement the primary mechanism for energy dissapation in uv absorbers. The nature of substituents on the aromatic ring would be expected to influence the extent to which this keto-enol isomerism contributes to photostabilization.

Selection of the best uv absorber for protection of a particular polymer must take into account several important factors other than the obvious efficiency-to-cost ratio. For example, o-hydroxybenzophenones are relatively inexpensive and provide excellent protection against the photodegradation of polyolefins, poly(vinyl chloride) and polyesters. These uv absorbers, however, may develop objectional color in some polymers during processing for exposure to sunlight. When color stability is essential, the more stable though more expensive, o-hydroxyphenylbenzotriazoles might be preferred. Also an antagonistic effect can occur with combinations of certain uv absorbers and antioxidants as has been observed with carbon black and amines or certain phenolic antioxidants (refer to Sect. D).

3) Radical Traps

Hindered amines are the latest category of photostabilizers to become available for the protection of polymers against uv degradation. Although the mechanism by which hindered amines protect against photodegradation is complex and has not as yet been completely established, radical trapping is generally agreed to be a major component of the process. Originally developed as photostabilizers for polyolefins, hindered amine light stabilizers (HALS) are now used to protect many other classes of polymers. There is considerable interest in these compounds because of their high level of efficiency at relatively low concentrations. At only 0.1 percent concentration, stabilizers in this category provide protection equivalent to 3 or 4 percent of a typical uv absorber. The best of these compounds is claimed to be as effective as carbon black in inhibiting photodegradation. Though they are expensive, HALS compounds provide the best performance in non-black formulations based on a cost-efficiency ratio.

HALS compounds based on 2,2,6,6-tetramethylpiperidine are attracting considerable attention [6-9]. Several important, commercial photostabilizers are now available based on this structure as typified by bis(2,2,6,6-tetramethyl-4-piperidinyl) sebacate (VIII).

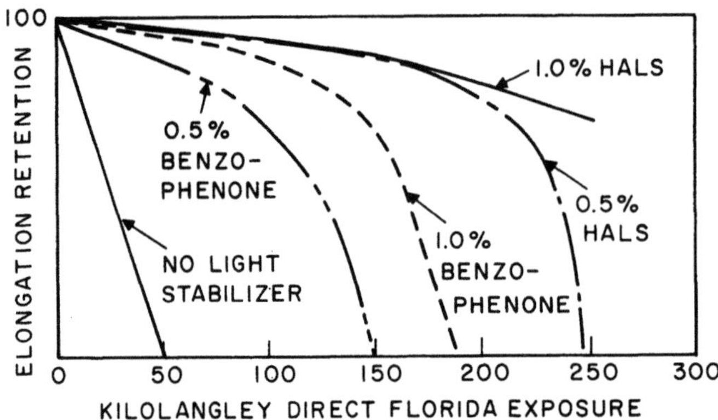

Fig. E-5. Elongation retention of light-stabilized high-density polyethylenes during outdoor exposure. (Reprinted with permission of The American Chemical Society)

The effectiveness of this stabilizer in protecting pigmented, high-density polyethylene during outdoor exposure is compared in Fig. E-5 with that obtained with an equal concentration of a typical uv absorber. Since HALS compounds do not absorb uv radiation above 270 nm [10, 11] they are usually not effective as uv absorbers, but must inhibit photodegradation by different reactions.

Several mechanisms have been suggested to account for photostabilization by hindered amines. Scott and coworkers [12] have proposed a mechanism involving reaction between HALS compounds and hydroperoxides, formed in the polymer, to yield stable nitroxyl radicals (IX) which are believed to be responsible for stabilization. These nitroxyl radicals, which apparently form in the early stages of reaction, are very effective scavengers of alkyl or macroalkyl radicals but not peroxy radicals [13]. Substituted hydroxylamines (X) are formed in this reaction.

It has also been suggested [14] that singlet oxygen can convert hindered amines into nitroxyl radicals.

It is evident that HALS compounds can inhibit polymer degradation in several different ways. The ability of hindered amines to react with hydroperoxides and of nitroxyl radicals to trap alkyl radicals suggests a role in protecting against thermal oxidation. The substituted hydroxylamines (X) can react with peroxy radicals, to regenerate nitroxyl radicals (IX). HALS compounds may also protect by deactivating excited states in polymer molecules [7].

$$\text{X} \quad \text{N–OH} + \text{ROO}^{\bullet} \longrightarrow \quad \text{IX} \quad \text{N–O}^{\bullet} + \text{ROOH}$$

For nitroxyl radicals to function effectively as radical traps, they must compete with oxygen in scavenging alkyl radicals. Wiles and associates [15] have found that the reaction of polymer radicals with nitroxyl radicals is about twenty times slower than with oxygen. However, regeneration of nitroxyl radicals as shown above could alter this ratio. Substituted hydroxylamines, on the other hand, react with peroxy radicals at a rate about a hundred times faster than these radicals react with hydrogen at branch sites in polypropylene. These authors have concluded that although radical scavenging by nitroxyl radicals and by substituted hydroxylamines can account for the effectiveness of some HALS compounds, other reactions must also contribute to the photostabilization of polypropylene. They suggest that association of the stabilizer or its reaction products with oxidation sites in the polymer could be an important reaction in the stabilization of polypropylene.

There are additives other than the hindered amines that can function as radical traps. Other additives, which protect primarily by other mechanisms can also trap active radicals in a secondary or supplemental process. For example, Bickel and Kooyman [16] have shown that peroxy radicals that are involved in thermal oxidation are trapped by radicals formed from chain-breaking antioxidants,

Quinones are very effective radical traps and are often used to scavenge radicals responsible for the spontaneous polymerization of monomers such as styrene. Carbon black and many polynuclear hydrocarbons [17] also function as radical traps. Though not important commercially, stable dialkyl nitroxide radicals (XI) have been shown to be effective scavengers of radicals.

$$(CH_3)_3C{-}\overset{\overset{\textstyle O^{\bullet}}{|}}{N}{-}C(CH_3)_3 \, . \quad \text{XI}$$

Even transitory radicals formed from an additive may persist long enough to become involved in scavenging propagating radicals.

When conventional antioxidants are used in combination with a uv absorber to protect against photodegradation, radical trapping by antioxidant products may contribute, to a minor extent at least, in the overall process of stabilization. In black formulations, some of the active radicals may be trapped by carbon black. Hindered amines, on the other hand, probably function primarily as radical traps, nitroxyl radicals and substituted hydroxylamines being the actual radical scavengers.

4) Quenchers

As discussed in the preceding sections, polymers can be protected effectively against photodegradation by providing a physical screen between the polymer and the light source, by additives that preferentially absorb damaging radiation, or by trapping radicals formed as the polymer degrades. Quenchers [18-22] inhibit photodegradation by yet another mechanism. These photostabilizers function by deactivating chromophores in excited polymer molecules through an energy-transfer mechanism before these excited states can undergo reactions that would result in degradation,

$$\left[\begin{array}{c} O \\ \| \\ R-C-R \end{array}\right]^{\circ} \xrightarrow{uv} \left[\begin{array}{c} O \\ \| \\ R-C-R \end{array}\right]^{*} \xrightarrow{Quencher^{\circ}} \left[\begin{array}{c} O \\ \| \\ R-C-R \end{array}\right]^{\circ} + Quencher^{*}$$

$$Quencher^{*} \longrightarrow Quencher^{\circ} + Heat$$

Where * is an excited state and ° is the ground state. Alternatively, an excited-state complex may be formed and this complex could then dissipate absorbed energy as heat or by photophysical processes.

$$\left[\begin{array}{c} O \\ \| \\ R-C-R \end{array}\right]^{\circ} + Quencher^{\circ} \longrightarrow \left[\begin{array}{c} O \\ \| \\ R-C-R---Quencher \end{array}\right]^{*} \longrightarrow Heat$$

Like the hindered amines, quenchers usually do not absorb uv radiation which would be damaging to the polymer. The primary role of quenchers is to extract absorbed energy from excited polymer molecules before these excited molecules are degraded. Chelates of transition metals form the most common class of quenchers. These compounds do not migrate through the polymer mass to a significant extent. Because of this low mobility, metal chelates are very effective stabilizers in thin cross sections or in fibers. Other types of photostabilizers such as the uv absorbers work well in thick sections since they migrate from the polymer bulk to the surface, continually replenishing the stabilizer in the area where photodegradation occurs. Metal chelates are very effective since they only react with polymer molecules which are in the excited state and are about to undergo degradation.

XII

Table E-4. Relationship between triplet quenching and ultraviolet stabilizing efficiency of metal chelates [23]

Stabilizer (based on formula XII)	Quenching effect on triplet anthracene	Hours required for polypropylene to absorb 0.06% O_2
R = OH	Strong	2,000
R = n-C_4H_8	Moderate	825
R = C_6H_5	None	470
2-Hydroxy-4-n-octyl-oxybenzophenone	None	650

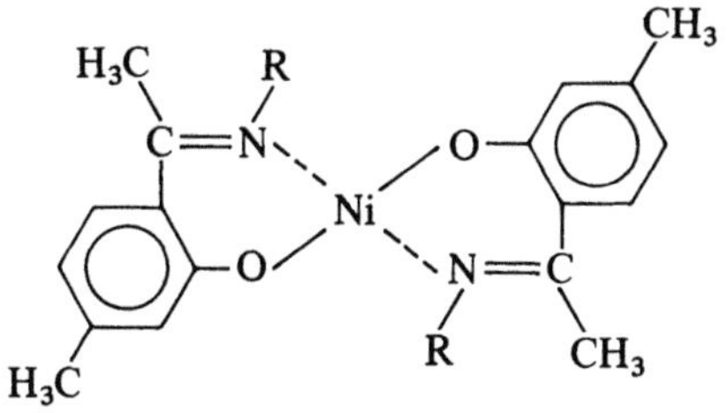

XII

Chelates of zinc and other transition metals (XII) deactivate photosensitized triplet states of anthracene [23], diethyl ketone [24], and benzophenone [25]. Triplet states are probably the excited species in most polymers as they undergo photodegradation. However, it has been suggested [26, 27] that quenching of excited-singlet states of carbonyl groups in some polymers may also be involved in photostabilization by metal chelates. Table E-4 shows the relationship between the quenching of triplet anthracene and the photostabilization of polypropylene. Although chelates of transition metals are the most important class of quenching agents used to stabilize polymers, a number of other compounds are known to be effective triplet quenchers. Cyclooctadiene for example is an effective stabilizer for poly(phenylvinyl ketone), but less effective in ethylene-carbon monoxide copolymers [21].

5) Stabilizer Combinations

Many efficient stabilizers are available for the protection of polymers during outdoor exposure, but for long-term exposure or for highly sensitive polymers, combinations of more than one stabilizer may be necessary. An important example of an effective stabilizer combination consists of carbon black and antioxidants in the protection of polyolefins during outdoor weathering. Antioxidants are used in this example to inhibit thermal degradation resulting from the elevated temperatures generated in black formulations. Since carbon black can function both as a thermal and a photostabilizer, it could be considered to function as a stabilizer combination. However, much greater protection is obtained with synergistic combinations of carbon black with selected antioxidants [28] (refer to Sect. II-A).

Most antioxidants when used alone give very little if any protection against photodegradation. Carbon black is the one important exception. However, to the extent that hydroperoxides or peroxy radicals are involved in photodegradation, antioxidants may

Table E-5. Synergism between a UV absorber and a thermal antioxidant [4]

Additive (0.4 pph)		Time to embrittlement of low-density polyethylene (in hours)
UV Stabilizer	Antioxidant	
None	None	400
Octylphenyl Salicylate	None	1,600
None	Tri(nonylphenyl) Phosphite	1,800
Octylphenyl Salicylate	Tri(nonylphenyl) Phosphite	7,000
2-Hydroxy-4-n-octoxybenzophenone	None	2,000
None	Tri(nonylphenyl) Phosphite	1,000
2-Hydroxy-4-n-octoxybenzophenone	Tri(nonylphenyl) Phosphite	8,500

play a role in deactivating these reactive intermediates. Several combinations of uv absorbers with phosphite-based antioxidants, listed in Table E-5, show a synergistic effect in polyethylene during outdoor weathering [4].

Since hindered amines do not absorb uv radiation in the range that is damaging to polymers, they are often used in combination with a uv absorber. These combinations provide a dual mechanism for stabilization against photodegradation. Accumulation of radicals and hydroperoxides is restricted by the uv absorber. Those reactive products which escape this first step can then be controlled by reaction products formed from the HALS compound as previously described. These combinations may be synergistic.

Allen [29] has reported antagonistic effects between bis(2,2,6,6-tetramethyl-4-piperidinyl) sebacate and several thermal antioxidants during processing of polypropylene. This antagonism results from sensitization of polypropylene to thermal breakdown by the hindered piperidine stabilizer. Other polyolefins probably undergo similar reactions but to a variable extent. Sulfur-containing antioxidants [30, 31] and transition metal salts of dialkyldithiocarbamates [32] show a pronounced antagonism. Both of these antioxidant types are effective decomposers of hydroperoxides, and destruction of these reactive intermediates would be expected to decrease the concentration of nitroxyl radicals formed by reaction of hydroperoxides with the hindered amines. This reaction is essential for the hindered amines to function as radical scavengers in the inhibition of photodegradation. Chain-breaking antioxidants, although they are not effective hydroperoxide decomposers, do reduce the amount of hydroperoxides formed and probably reduce activity of the hindered amines in the same way.

6) Stabilization by Polymer Modification

The same principles of polymer modification to improve thermal stability (refer to Sect. D-I-6) apply to photostabilization. To the extent that labile or photosensitive groups can be replaced by less reactive groups in a polymer molecule, improved resistance to photodegradation will result. Thus polytetrafluoroethylene would be expected to be more stable than polyethylene, and it is. This approach to photostabilization is, however, rarely if ever practical since major changes such as replacing all C—H bonds with C—F bonds in a polymer obviously alters other properties beyond acceptable limits. Less drastic changes in structure, however, could provide practical routes to photostabilization.

Saturated hydrocarbons of low-molecular-weight have good light stability as contrasted to the light sensitivity of hydrocarbon polymers. As was pointed out in Sect. B-II-2, carbonyl groups in commercial hydrocarbon polymers are the primary chromophores responsible for the absorption of damaging uv radiation. In theory at least, removal or modification of these light-sensitive groups should lead to improved resistance to photodegradation. Such reactions are possible on a laboratory scale, but have not been used in production because of obvious economic considerations. However, reduction in the carbonyl content of commercial hydrocarbon polymers, which would lead to improved photostability could be achieved through more carefully controlled polymerization and processing techniques. This is evident in the wide variations observed in the carbonyl contents and light sensitivities of hydrocarbon polymers from various sources.

The incorporation of photostabilizers into polymer molecules by grafting or copolymerization, using comonomers having a stabilizing moiety in their structure, has been investigated[33] as a route to improved photostability, but this technology has yet to be used to a significant extent in commercial applications. Sensitive polymers can be stabilized, however, by laminating a light-stable polymer over the exposed surface. Those polymers having low compatibility with uv absorbers could be protected by a surface-laminated film of a second polymer capable of retaining a high concentration of the required additive. In this way, damaging radiation would be screened from the more vulnerable polymer. None of these techniques for improving photostability by polymer modification, however, is competitive with the customary procedure of blending selected photostabilizers into a polymer.

II) Stabilization Against Ionizing Radiation

Ionizing or high-energy radiation generates more complex reactions in polymers than are believed to take place during uv radiation. Ionizing radiation is produced by several sources, each capable of ionizing polymer molecules. X-rays, gamma rays and high-energy particles produced in nuclear reactors have been studied extensively as radiation sources in polymer degradation, and, perhaps more importantly, for the deliberate production of cross-links in polymers to improve or alter certain properties. Production of heat-shrinkable tubing is an example of a process in which cross-links are deliberately introduced into a polymer. Stabilization is of prime importance in designed cross-linking processes, and the effects of radiation on both polymer and stabilizer must be taken into account.

In ionizing radiation, energy absorption is not specific for any particular bond in the polymer, but depends instead on the electron density of all bonds in the structure. Oxygen is usually present during exposure to radiation and oxidative reactions can contribute to polymer failure. When radiation is used to generate cross-links, however, the process is usually run in the absence of oxygen. A different approach to stabilization should be used if oxidative reactions contribute to degradation.

Random, homolytic bond cleavage is the predominant reaction occurring in many polymers exposed to ionizing radiation, but competitive reactions of cross-linking often take place simultaneously,

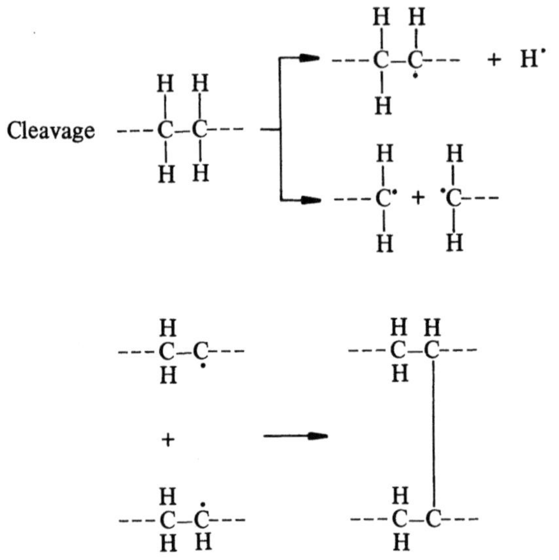

Molecular fragmentation can occur by cleavage of bonds along the backbone chain, but recombination of radicals produced by this reaction is highly probable due to the cage effect [34]. Cleavage of C—H bonds predominates in those polymers having only hydrogen attached to the backbone chain, despite the fact that these bonds have a higher bond dissociation energy than C—C bonds. Also radicals are produced in clusters during ionizing radiation and this favors combination of hydrogen atoms to form molecular hydrogen which then escapes from the polymer matrix.

Radiation mechanisms are further complicated by ionic reactions which are believed to occur simultaneously with radical reactions [35, 36]. Ion-molecule reactions may also contribute to degradation by ionizing radiation [35, 37]. The multiplicity of mechanisms believed to be involved in radiation-induced degradation complicates consideration of stabilization. Four general approaches have evolved for the stabilization of polymers against ionizing radiation:

1. Scavenging of radicals formed from excited polymer molecules to yield inert reaction products. Additives that protect by this mechanism perform a sacrificial role and hence are unlikely to provide long-term protection.
2. Energy transfer from excited polymer molecules to an additive which then dissipates the energy without further reaction occurring in the polymer.
3. A compensating process in which the approximate molecular weight of the degraded polymer is restored.
4. Modification of the polymer structure to form a polymer having improved resistance to radiation.

   Though some of these approaches are similar to those described in stabilization against other modes of degradation, there are sufficient differences to warrant a more detailed discussion.

Radical scavengers or antirads function as hydrogen donors to repair damage due to the loss of hydrogen atoms.

$$PH \xrightarrow[\text{Radiation}]{\text{Ionizing}} \begin{cases} P^\bullet \xrightarrow{\text{IIA}} PH + A^\bullet \\ PH^+ \xrightarrow{\text{A}} PH + {}^\bullet A^+ . \end{cases}$$

Where PH is a polymer molecule.

The first of these reactions is identical with that described in stabilization against thermal oxidation, and antioxidants can indeed also function as antirads [38, 39]. In contrast to thermal stabilization, however, when these hydrogen donors function as antirads, they apparently bond to the polymer since they cannot be extracted after reaction has occurred [40]. Thus the reaction of antioxidants as antirads appears to be more complex than the simple process of donating labile hydrogen to a radical. The level of protection increases with increasing concentration of antirads. Even polyolefins, normally quite vulnerable to ionizing radiation, can withstand very high doses of radiation when protected by an effective antirad used in adequate concentration.

Energy transfer reactions as a route to stabilization against ionizing radiation have not been studied as extensively as radical scavenging reactions. By analogy to photostabilization mechanisms, however, it can be postulated that quenchers having an aromatic structure would protect against radiation damage [34]. Aromatic ring structures provide a route to the harmless dissipation of energy transferred from excited polymer molecules as will be discussed under stabilization by structure modification.

Stabilization by compensating reactions is not true stabilization since the degraded polymer is not restored to its original structure. In polymers that cross-link during exposure to radiation, scission of polymer chains can occur simultaneously, resulting in a final molecular weight approximating that of the original polymer. Conversely, in polymers that undergo chain scission, cross-linking could be enhanced by additives to approach the original molecular weight [34].

Protection by structure modification results when aromatic moieties are incorporated into polymers since aromatic rings are normally sensitive to radiation. This effect is shown in Table E-6 for a series of copolymers of isobutylene and styrene with varying styrene content [41]. Apparently, the aromatic ring in each styrene unit inhibits oxidation in styrene units and in addition protects one or two neighboring isobutylene units. Acrylonitrile-styrene copolymers are also resistant to ionizing radiation [42], but in these polymers the maximum protection occurs at about 0.05 mole percent fraction of styrene. Several additional examples of copolymers with good resistance to radiative damage resulting from aromatic rings in their structure have been reported [43–45].

Table E-6. Stability of isobutylene-styrene copolymers to ionizing radiation [41]

Mole percent of styrene	No. of chain scissions per 100 electron volts
0	5.9
20	3.0
50	1.8
80	1.0

References

1. Hirt, R.C., Searle, N.Z.: Energy Characteristics of Outdoor and Indoor Exposure Sources and Their Relation to the Weatherability of Plastics, in: Weatherability of Plastic Materials (ed) Kamal, M.R., p. 61, New York: Interscience 1967
2. Hawkins, W.L., Winslow, F.H.: Antioxidant Properties of Carbon Black, in: Reinforcement of Elastomers (ed) Kraus, G., p. 565, New York: Interscience 1965
3. Wallder, V.T. et al.: Ind. and Eng. Chem., *42*, 2320 (1950)
4. Chaudet, J.H. et al.: Trans. Soc. Plastics Engs., *1*, 26 (1961)
5. Tobin, W., Vigeant, F.: Plastics Compounding, Sept./Oct. 1961, p. 16
6. Patel, A.R., Usilton, J.J.: Ultraviolet Stabilization of Polymers: Development with Hindered-Amine Light Stabilizers, Advances in Chemistry Series, Amer. Chem. Soc., *169*, 116 (1978)
7. Carlsson, D.J., Wiles, D.M.: J. Macromol. Sci., Rev. Macromol. Chem., *C14*, 155 (1976)
8. Gugumus, F.: Developments in the U. V. Stabilization of Polymers, in: Developments in Polymer Stabilization (ed) Scott, G., p. 261, Applied Science Publishers, London 1979
9. Allen, N.S., McKellar, J.F.: Brit. Polym. J., *9*, 302 (1977)
10. Heller, H.J., Blattman, H.R.: Pure and Appl. Chem., *36*, 141 (1973)
11. Felder, B., Schumacher, R.: Macromol. Chem., *31*, 35 (1973)
12. Chakraborty, K.B., Scott, G.: Polymer, *18*, 99 (1977)
13. Bolsman, T.A.B.M. et al.: Rec. Trav. Chim. Pays-Bas, *97*, 313 (1978)
14. Ivanov, V.B. et al.: Photochem., *4*, 313 (1965)
15. Grattan, D.W., Carlsson, D.J., Wiles, D.M.: Polym. Degrad. and Stab.: *1 No. 1*, 69 (1979)
16. Bickel, A.F., Kooyman, E.C.: J. Chem. Soc. (London), 3211 (1953); 2215 (1956); 2217 (1957)
17. Ingold, K.U.: Inhibition of Autoxidation, Advances in Chemistry Series, Amer. Chem. Soc., *85*, 296 (1968)
18. Melchore, J.A.: Ind. Eng. Chem. Prod. Res. *1*, 232 (1963)
19. Schmitt, R.G., Hirt, R.C.: J. Appl. Polym. Sci., *7*, 1565 (1963)
20. Hormann, H.H.: Ind. Eng. Chem. Prod. Res. Rev., *5*, 92 (1966)
21. Heskins, M., Guillet, J.: Macromol., *3*, 224 (1970)
22. Trozzolo, A.M.: Stabilization Against Oxidative Photodegradation, in: Polymer Stabilization (ed) Hawkins, W.L., p. 199, New York: Wiley-Interscience 1972
23. Briggs, P.J., McKellar, J.F.: J. Appl. Polym. Sci., *12*, 1825 (1968)
24. Chien, J.C.W., Conner, W.P.: J. Amer. Chem. Soc., *90*, 1001 (1968)
25. Harper, D.J., McKellar, J.F., Turner, P.H.: J. Appl. Polym. Sci., *18*, 2805 (1974)
26. Coulson, D.R., Yang, N.C.: J. Amer. Chem. Soc., *88*, 4511 (1966)
27. Wagner, P.J., Hammond, G.S.: J. Amer. Chem. Soc., *88*, 1245 (1966)
28. Hawkins, W.L., Worthington, M.A.: J. Polym. Sci., *Part A 1*, 3489 (1963)
29. Allen, N.S.: Polym. Degrad. and Stab., *3 (No. 1)*, 73 (1980)
30. Chakraborty, K.B., Scott, G.: Chem. and Ind. (London), 237 (1978)
31. Allen, N.S., McKellar, J.F., Wilson, D.: Chem. and Ind. (London), 887 (1978)
32. Tozzi, A., Cantatore, G., Masina, F.: Text. Res. J., *48*, 433 (1978)
33. Bailey, D., Vogel, O.: J. Macromol. Sci., *C14*(2), 267 (1976)
34. Lyons, B.J., Lanza, V.L.: Protection Against Ionizing Radiation, in: Polymer Stabilization (ed) Hawkins, W.L., p. 249, New York: Wiley-Interscience 1972
35. Collinson, E., Dainton, F.S., Walker, D.C.: Trans. Faraday Soc., *57*, 1732 (1961)
36. Borovkova, V.A., Bagdasar'yan, Kh.S.: Khim. Vys. Energ., *1*, 340 (1967)
37. Stevenson, D.P.: J. Phys. Chem., *61*, 1453 (1957)
38. Charlesby, A.: Atomic Radiation in Polymers, New York: Pergamon Press 1960
39. Charlesby, A., Lloyd, D.G.: Proc. Roy. Soc. (London), *254A*, 343 (1958)
40. Lanza, V.L.: Irradiation Property Changes, in: Crystalline Olefin Polymers (eds) Raff, R.A.V., Doak, K.W., p. 337, New York: Wiley-Interscience 1964
41. Charlesby, A., Alexander, P.: J. Chem. Phys., *52*, 699 (1955)
42. Weber, E.J., Hensinger, H.: Radiochim. Acta, *4*, 92 (1965)
43. Gardner, J.B., Harper, B.G.: J. Appl. Polym. Sci., *9*, 1585 (1965)
44. Schnabel, W.: Makromol. Chem., *104*, 1 (1967)
45. Arthur, J.C., Jr. et al.: J. Appl. Polym. Sci., *11*, 1129 (1967)

F) Stabilization Against Degradation by Ozone

The reactions responsible for ozone-induced degradation of elastomers under stress were discussed briefly in Sect. B-II-3. Although this form of degradation is limited to unsaturated polymers, it is nonetheless of considerable importance. Degradation in the presence of ozone was first observed in natural rubber, but it is now known to take place in many synthetic elastomers as well. The general reaction is one of cleavage involving ethylenic double bonds in the backbone chain. Stabilization of natural rubber has been accomplished by both physical and chemical methods, and by a combination of the two approaches. Synthetic elastomers can also be protected against ozone-induced degradation using the same technology.

I) Stabilization by Waxes

The addition of waxes to compositions of natural rubber to protect against ozone had been used for many years [1], long before the chemistry of this degradation was understood. Waxes protect by a physical process in which the additive blooms out of the formulation to form a protective film over the surface. The wax film is relatively unreactive toward ozone. As long as it remains intact, the elastomer is protected against degradation. However, waxes function most effectively in static applications. When the elastomer is to be used in dynamic applications where flexing occurs, the protective films may flake off leaving the elastomer exposed to ozone attack.

Degradation by ozone is evident as surface cracks, always perpendicular to the direction of applied stress. Under biaxial stress, a checkerboard pattern is observed. Surface cracks may eventually propagate through the material leading eventually to fracture. The first type of failure resulting from ozone-induced degradation is usually due to deterioration of surface appearance. However, as the reaction proceeds, mechanical failure takes place.

The amount of wax used to protect against ozone attack is important. Sufficient wax must be used to form a continuous film over the stressed elastomer [2], but an excess tends to accelerate the flaking off process. Wrapping automobile tires with wax impregnated paper has been used in areas of high ozone concentration to preserve the surface appearance.

Both paraffin and microcrystalline waxes are used in the stabilization of stressed elastomers. Both types of waxes are obtained from petroleum, but the microcrystalline waxes have a higher average molecular weight. Paraffin waxes are composed of hydrocarbons ranging from $C_{18}H_{38}$ to $C_{32}H_{66}$. Paraffin waxes bloom rapidly to the surface but flake off rather easily. Microcrystalline waxes, on the other hand, bloom more slowly[3] but form a more cohesive protective layer that resists flaking off. Optimum protection is obtained by using combinations of a paraffin and a microcrystalline wax. In these combinations, the paraffin component accelerates migration of the microcrystalline wax to the surface. An effective film of the two waxes then forms rapidly enough to provide effective protection.

Choice of specific waxes of each general type is still an arbitrary decision[4] although relationships have been suggested between efficiency in protection and certain physical properties of the wax[3]. Conditions of exposure to ozone may also influence the level of protection provided by a particular wax[5]. The high concentrations of ozone used in accelerated testing may indicate poor stability unless adequate time is allowed for the wax film to form at the surface. When dynamic conditions are involved, combinations of waxes and chemical stabilizers (antiozonants) are usually required for adequate protection.

II) Stabilization by Antiozonants

In order to understand how antiozonants protect stressed elastomers against ozone attack, it is first necessary to consider in more detail the reactions of ozone with ethylenic double bonds. The eventual result of these reactions is cleavage of the double bonds and the overall reaction is referred to as ozonlysis. The reaction of low-molecular-weight olefins with ozone has been studied extensively by many investigators[6-9].

Huisgen[10] suggested that the initial reaction product of ozone with a double bond is a 1,2,3-trioxolane (I).

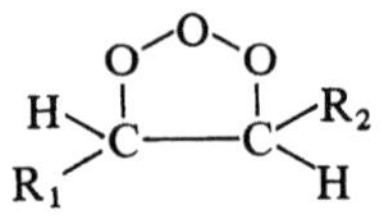

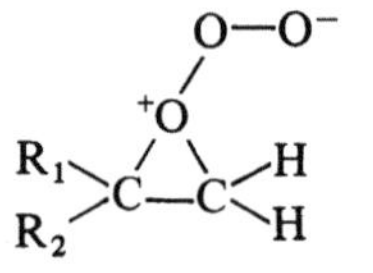

A one step addition to form this initial product has been suggested[11]. One of the C—C bonds in the olefin remains intact in the initial reaction. Many adducts of this type have been isolated in the reaction of trans olefins with ozone[8,12], but not from the corresponding cis isomers. Olefins having terminal unsaturation may form a different initial adduct (II)[13,14]. This structure is supported by the high yield of epoxy products from these olefins[13-15]. Criegee[16] has proposed a mechanism to account for the scission of ethylenic bonds by ozone. Although this mechanism is based on model compound studies, it does afford a sound basis for discussion of the ozone-induced degradation of stressed elastomers.

The 1,2,3-trioxolane (I), assumed to be the initial reaction product in Criegee's mechanism, decomposes to form a zwitterion (III) and a carbonyl compound (IV).

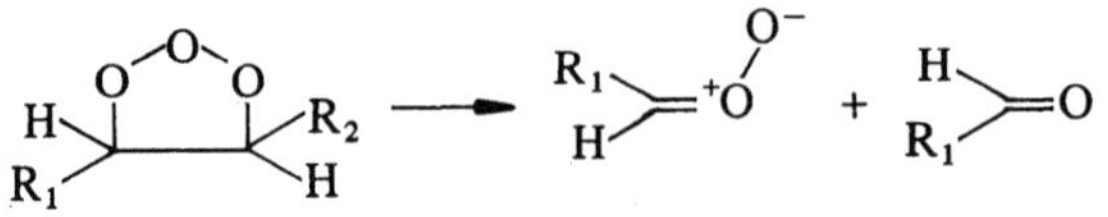

The zwitterion may then undergo several different reactions. Recombination with the carbonyl compound, if the latter is sufficiently reactive, would yield an ozonide (V) [1]. The remaining C—C bond in the original olefin is cleaved when the trioxalane decomposes. This is also the case when recombination occurs to form an ozonide.

Unsymmetrical olefins should produce two different zwitterions and two different carbonyl compounds,

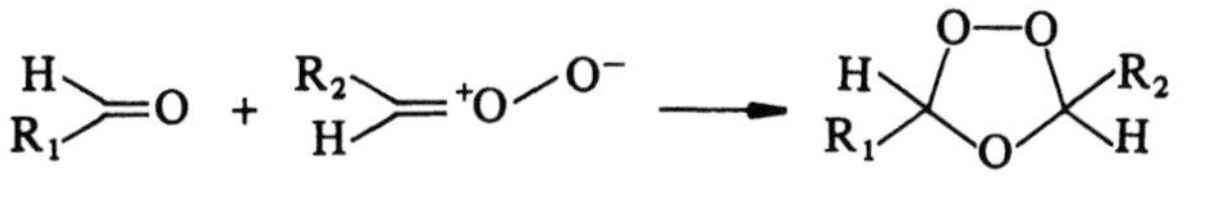

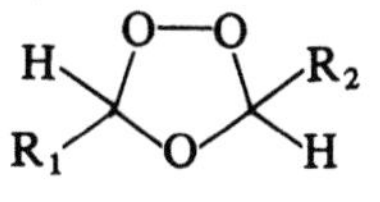

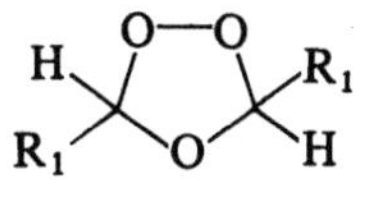

Random recombination of these four products should then yield the following three ozonides (V, VI, VII), each of which could exist as a cis-trans pair. Several investigators [18-20], using carefully selected reaction conditions, have succeeded in isolating all six products from simple olefins. However, substantially different cis-trans ratios of the ozonides have been reported [21] in the ozonolysis of cis- and trans-4-methyl-2-pentene. This observation suggests that the simple recombination of a zwitterion and a carbonyl compound, as proposed by Criegee, cannot completely explain this complex series of reactions.

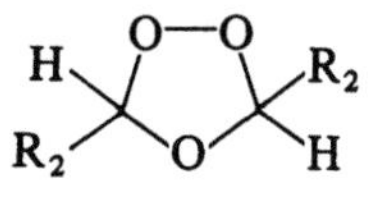

Alternatively, the zwitterion may dimerize to a diperoxide (VIII) or polymerize to a polymeric peroxide (IX).

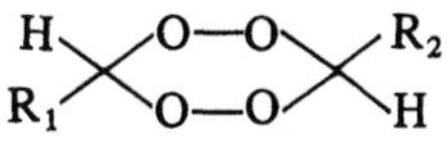

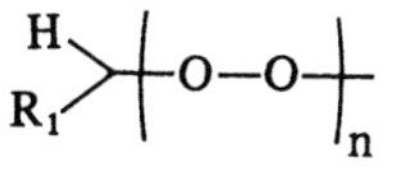

Polymeric ozonides could also form from a more random recombination of a zwitterion and a carbonyl compound. Recombination to form the six isomeric ozonides, however, is believed to be a principal reaction in most instances.

The key point to be derived from this review of the Criegee mechanism is that several intermediate products are formed during the ozonolysis of olefins. These intermediates must certainly be involved in stabilization reactions with antiozonants. It should not be surprising therefore that the mechanism by which antiozonants protect against ozone attack is quite complex. Several interesting mechanisms have been proposed to explain stabilization by antiozonants, and these will be discussed with the evidence available to support each of them.

The structure of many antiozonants is similar to that of the N,N'-diaryl-p-phenylenediamines used as antioxidants to inhibit thermal oxidation. In contrast to these antioxidants, however, effective antiozonants, based on the p-phenylenediamine structure, have alkyl rather than aryl substituents. Conventional antioxidants provide very little protection against ozone attack. Similarly, those N,N'-dialkyl-p-phenylenediamines which are effective as antiozonants give little protection against thermal oxida-

F) Stabilization Against Degradation by Ozone

Table F-1. Typical antiozonant structures

Staining:

Nonstaining:

Note: All R groups are alkyl

tion. It should also be recalled that antioxidants can function in the polymer bulk to the extent that oxygen reaches these areas. Antiozonants, on the other hand, react only at the surface where reaction with ozone takes place. Structures for representative antiozonants are shown in Table F-1. These antiozonants are classified as staining or nonstaining. In general, those antiozonants based on the p-phenylenediamine structure yield diimine products which are intense chromophores. Nonstaining antiozonants, on the other hand, do not form diimines and hence cause less discoloration.

The so-called scavenger theory has found considerable support[22-24] since it is a simple explanation for the process by which antiozonants protect against ozone-induced degradation. This theory is based on the high reactivity of ozone with most organic compounds. It assumes that antiozonants at the surface react with and destroy ozone before it can reach the stressed elastomer. The scavenging of ozone by antiozonants, however, could only be effective if there were a sufficient concentration at the surface to react with all of the ozone reaching the elastomer. It is possible that antiozonants could diffuse from within the elastomer to the surface thus replenishing the protective film, but diffusion of these additives has been found to be much slower than would be required for this to occur [25]. Furthermore, according to this theory, the effectiveness of an antiozonant should improve with increasing concentration, but this is not generally observed [26]. Although there is considerable evidence against the scavenger theory as a complete or even the principal mechanism, preferential reaction of an antiozonant with ozone may still contribute to the overall stabilization process.

A second proposed mechanism suggests that as an antiozonant blooms to the surface it forms an impermeable barrier which prevents ozone from reaching the stressed elastomer [27, 28]. This explanation would require a low level of reactivity of the antiozonant with ozone that would destroy the protective film. There is abundant evidence to refute this conclusion. Reaction products of the antiozonant with ozone could constitute the protective film, but once again a more rapid diffusion of antiozonant to the surface would be required than has been observed [25].

As suggested previously, intermediate products from the ozonolysis reaction could be expected to enter into the mechanism for protection by antiozonants. It has been reported [29] that antiozonants react with ozonides, and with peroxidic and aldehydic products formed during ozonolysis. This has led to the development of the recombination theory in which bifunctional dialkyl-p-phenylenediamines are assumed to bridge over ruptured C—C bonds to in essence restore the backbone chain of the elastomer. In support of this mechanism, it has been found [29] that part of the added antiozonant is bonded to the elastomer network. Further support is found in observations [30] that effective antiozonants increase the stress required for cracking to take place. This could be the result of a stronger bond structure in the antiozonant bridge, but increased flexibility in the reconstructed bridge is a more likely explanation. The low diffusion rate of antiozonants through an elastomer, however, is not compatible with this mechanism. It is doubtful that sufficient antioxidant could reach the surface for the recombination to take place on initial exposure or during continued exposure to ozone.

Murray [31] has proposed another mechanism for the protection by antiozonants which finds its basis in the reduction or elimination of the rigid structure formed at the surface by ozonides. Interception of zwitterions by amine antiozonants (R_2NH) would prevent formation of rigid ozonide structures.

As a result of this reaction, the surface would become more relaxed and mobile. Although bonds along the backbone chain would be ruptured, cracking would be suppressed as the surface becomes more relaxed. Diamine antiozonants may also decompose ozonides by a nucleohilic displacement at the peroxide bond.

Amines react rapidly with ozonides in model olefins[31], and it has already been mentioned that a portion of the antiozonant becomes attached to elastomer molecules[29]. Also according to this mechanism the most effective antiozonants should be those having the greatest nucleophilicity as was observed by Furukawa[32]. These and other observations lend strong support to this mechanism which proposes that the active stabilizer is not the antiozonant but rather it is a reaction product, formed in situ as a combination of ozonized elastomer and antiozonant. Sufficient time must be allowed for the suggested reactions to occur before effective protection is realized. Thus exposure of an elastomer formulation, containing an antiozonant, to high initial concentrations of ozone, as would be experienced in an accelerated test, could show little or no protection. This same formulation, however, might show excellent stability to ozone at the low concentrations that would be present during normal, outdoor exposure.

III) Stabilization by Structure Modification

Modification of the surface of an elastomer can provide improved resistance to degradation by ozone. Elimination of double bonds at the immediate surface has been used as a structural modification[33] to improve ozone resistance. Limited oxidation of the surface of natural rubber has been suggested[34] as another approach to improving stability against ozone attack. Cross-linking at the surface provides greater resistance to cracking. This surface modification has been obtained by exposure to heat and to photooxidation, but these oxidative reactions may produce a brittle surface which could be vulnerable to stress cracking.

Copolymerization can provide a modified elastomer with improved resistance to ozone, but often with a loss in elastomeric properties. Blending of acrylonitrile-styrene elastomers with poly(vinyl chloride) has been reported by several investigators[35-37] to yield ozone-resistant products. A variety of continuous coatings that are inert to ozone have been used to protect natural rubber against ozone attack[38,39]. These protective coatings function much like the films of wax which bloom to the surface of elastomers, but have the advantage that they give immediate protection. Properly selected coatings may have the additional advantage over waxes of greater adhesion to the elastomer surface.

References

1. Kreusler and Budde: German Patent No. 18 740, August 26, 1881
2. Braden, M.: J. Appl. Polym. Chem., *3*, 100 (1960)
3. Winkelman, H.A.: Ind. Eng. Chem., *44*, 841 (1952)
4. Ferris, S.W., Kurtz, S.S., Jr., Sweely, J.S.: Amer. Soc. Testing Matls., Spec. Tech. Publ., *No. 229*, 72 (1958)
5. Crabtree, J., Kemp, A.R.: Ind. Eng. Chem., Anal. Ed., *18*, 769 (1946)
6. Criegee, R.: Rec. Chem. Progr. (Kresge-Hooker Sci. Lib.), *18*, 111 (1957)
7. Wibaut, J.P. et al.: Rec. Trav. Chim. Pays-Bas, *71*, 761 (1952)
8. Greenwald, F.L.: J. Org. Chem., *30*, 3108 (1965)
9. Bailey, P.S., Thompson, J.A., Shoulders, B.A.: J. Amer. Chem. Soc., *88*, 4098 (1966)
10. Huisgen, R.: Angew. Chem., Int. Ed., *2*, 565 (1963)
11. Criegee, R., Schröder, G.: Chem. Ber., *93*, 689 (1960)

12. Murray, R.W., Youssefyeh, R.D., Story, P.R.: J. Amer. Chem. Soc., *89*, 2429 (1967)
13. Bailey, P.S., Lane, A.G.: J. Amer. Chem. Soc., *89*, 4473 (1967)
14. Bartlett, P.D., Stiles, M.: J. Amer. Chem. Soc., *77*, 2806 (1955)
15. Criegee, R.: Advances in Chemistry Series, Amer. Chem. Soc., *No. 21*, 133 (1959)
16. Criegee, R.: Peroxide Reaction Mechanisms (ed) Edwards, J.O., p. 29, New York: Interscience 1962
17. Criegee, R., Bath, S.S., Bornhaupt, B.V.: Chem. Ber., *93*, 2891 (1960)
18. Riezebos, G., Grimmelikhuysen, J.C., Van Dorp, D.A.: Rec. Trav. Chim. Pays-Bas, *82*, 1234 (1963)
19. Privett, O.S., Nickell, E.C.: J. Amer. Chem. Soc., *40*, 22 (1963)
20. Loan, L.D., Murray, R.W., Story, P.R.: J. Amer. Chem. Soc., *87*, 737 (1965)
21. Murray, R.W., Youssefyeh, R.D., Story, P.R.: J. Amer. Chem. Soc., *88*, 3143 (1966)
22. Biggs, B.S.: Rubber Chem. Technol., *31*, 1015 (1958)
23. Sullivan, F.A.V., Davis, A.R.: Rubber World, *141*, 240 (1959)
24. England, W.D., Krimian, J.A., Heinrich, R.H.: Rubber Chem. Technol., *32*, 899 (1960)
25. Braden, M.: J. Appl. Polym. Sci., *6*, 66 (1962)
26. van Pul, B.I.C.F.: Rubber Chem. Technol., *31*, 1117 (1959)
27. Murray, R.M.: Rubber Chem. Technol., *31*, 882 (1958)
28. Erickson, E.R. et al.: Amer. Soc. Testing Matls., Spec. Publ. *No. 229*, 11 (1959)
29. Lorenz, O., Parks, C.R.: Rubber Chem. Technol., *36*, 201 (1963)
30. Braden, M., Gent, A.N.: J. Appl. Polym. Sci., *6*, 449 (1962)
31. Murray, R.W.: Prevention of Degradation by Ozone, in: Polymer Stabilization (ed) Hawkins, W.L., p. 215, New York: Wiley-Interscience 1972

G) Test Procedures

It is important to determine or make a reasonable estimate of the useful life of a polymer before its degradation exceeds tolerable limits. Many tests have been devised to measure the rate of polymer degradation or the time to failure of polymer compositions. Methods used for testing are quite varied, reflecting the anticipated exposure conditions. Test procedures are conveniently classified as weatherability tests designed to evaluate out-door stability and thermal tests which measure the resistance to heat in the absence of radiation. Thermal tests are particularly useful in determining the extent of degradation which may occur under the high-temperatures encountered during processing. They are also used to estimate the useful life of polymers that will be exposed to elevated temperatures in the absence of solar radiation for long periods of time at or near ambient temperatures. Special test procedures have also been developed to measure ozone resistance, biodegradability, flammability, and the resistance to solvents or nonsolvents.

In designing test procedures, it is important to establish that specific property which will be responsible for failure. In many applications, polymers fail because of unacceptable changes in appearance as for example discoloration or the appearance of surface cracking. Failure in aesthetic properties, since appearance is a surface effect, can occur before there is a significant loss in mechanical strength. If a polymer is to be used in applications that depend primarily on mechanical properties, limited deterioration in aesthetic properties may be tolerable. Certain electrical properties, e.g. surface resistivity, are very sensitive to degradation at the polymer surface. Failure, as judged by these electrical properties, may also take place before there is any significant loss in mechanical strength.

I) Methodology

Tests for measuring the time to failure of polymers are classified as design or materials tests. Design tests measure the useful life of polymers as they function in an actual design or device. This type of test provides the ultimate answer for the design engineer. Design tests take into account not only the external environment but also every other component of the final product that could have an adverse (or beneficial) effect on a polymer's stability.

In contrast, materials tests evaluate only the stability of test polymer samples without consideration of how the polymer samples may perform in a final product. Materials tests are useful in the development and rating of stabilizers. A polymer may show acceptable stability in a materials test but fail a design test. This is exemplified by the rapid failure of stabilized polymers when they are used in contact with active metals. (Metal-catalyzed oxidation was discussed in Sect. D-I-3 a.) In other applications, stabilized polymers may be used in contact with solvents that extract stabilizers, leading to failure of a formulation that would show adequate stability in a materials test.

Accelerated testing is essential for estimating the useful life of most stabilized polymers or even for those unmodified polymers that have inherent stability due to their molecular structure. Degradation is normally accelerated by intensifying exposure conditions. For example, thermal stability is measured at temperatures substantially higher than would be encountered under conditions of actual use. The need for accelerated tests is demonstrated by the testing of stabilized polyethylene. When properly stabilized, this polymer resists thermal oxidation for many years at or near ambient temperatures. Obviously it would not be practical to test these formulations under normal exposure conditions. Even when tested at 140 °C, stabilized polyethylene undergoes negligible thermal oxidation for hundreds of hours. Accelerated testing at several elevated temperatures must be used. The data obtained under these conditions are then extrapolated to the temperature range expected to be encountered in use.

Extrapolation of accelerated test data to predict the useful life of a polymer, however, can lead to erroneous conclusions. An accurate prediction of stability to oxidative degradation can only be made if there is an established relationship between stability and temperature over the entire range of testing and use temperatures. Often the extrapolation deviates from linearity, though to varying degrees as suggested in Fig. G-1. If the relationship is nearly linear as in Curve A, a reasonable prediction of the time to failure is possible. In the relationship indicated by Curve B, the polymer would fail much sooner than indicated by the accelerated test data. Curve C is an example of a polymer formulation that is more resistant to degradation under conditions of actual use longer than would be expected from accelerated test data. The stability of carbon black/polyethylene formulations (refer to Fig. D-11) is a classical example of a polymer formulation exhibiting this type of relationship.

Divergence from linearity in extrapolating test data may be due to different rate constants for degradation reactions in the high and low temperature regions. Mechanisms may be different under accelerated conditions from those contributing to low-temperature degradation. Even a minor difference in reaction rates could contribute to divergence from linearity. A change in morphology is believed to be responsible for the pronounced nonlinearity observed in extrapolating accelerated test data obtained with carbon black/polyethylene formulations. Because of these and other factors which can modify the linear relationship, mathematical calculations usually cannot be used with confidence. If the rate of heating rather than the actual test temperature is the primary cause of divergence from linearity, an Arrhenius calculation might be used [1]. However, in most instances the test temperature rather than heating rate appears to be the important variable.

The most important conclusion to be drawn from the preceding discussion is that the closer test conditions approach those of actual use, the more reliable will be the prediction of service life.

There is another, quite different approach to accelerated testing which could give a more accurate prediction of the useful life of a polymer. This approach, referred to as early detection, is based on the capability to detect and measure accurately changes as they occur early in the degradation process. If such measurements were fully developed,

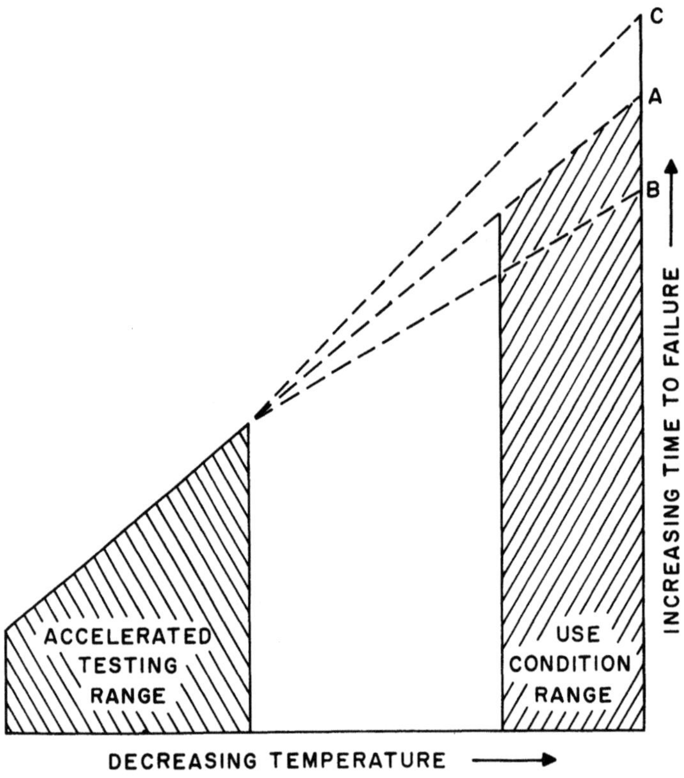

Fig. G-1. Extrapolation of time to failure from accelerated testing data

the polymer could then be exposed to conditions at or near those of actual use. The progress of incipient degradation would then be followed. Sensitive electrical tests, instrumental analysis [2-7] and chemiluminescence [8] have been considered as possible techniques which could be adaptable to the early detection approach. However, wide acceptance of early detection as a means of predicting the useful life of polymers has not as yet taken place.

Accelerated tests for weatherability are widely used to determine the outdoor life of polymers. Exposure at geographic locations where there is a high average level of ultraviolet (uv) radiation provides a small degree of acceleration. Indoor testing for weatherability using intense sources of uv radiation with continuous exposure provides a much greater degree of acceleration, but these tests do not always correctly simulate outdoor weathering conditions.

II) Properties Used to Evaluate Degradation

The degradation of polymers can be followed by many, different analytical techniques. Since most types of degradation result from irreversible chemical reactions of polymer molecules, changes in molecular composition provide a logical method for measuring the course of polymer degradation. Changes in chemical composition which take place during degradation vary with different polymers. The loss of hydrogen chloride provides a simple though sensitive method for following the degradation of poly(vinyl chloride).

As hydrocarbon polymers degrade, many different changes in polymer composition occur. The rate of carbonyl formation can be used to follow oxidative degradation in some polymers. Other groups including hydroperoxides and ethylenic groups also accumulate during oxidative degradation. However, the rate at which these groups form varies as chemical reactions proceed. Hydroperoxides are generally agreed to be initial degradation products, but these important intermediates also decompose as degradation proceeds. This is evident in the relationship between the rate of oxidation and hydroperoxide concentration as shown in Fig. B-2. Hydroperoxide accumulation can therefore only be accurate in the earliest stages of degradation. Carbonyl formation, on the other hand, continues at a fairly constant rate over a much longer period, and analysis of carbonyl content is a popular method for following the oxidative degradation of hydrocarbon polymers.

The effects of degradation on polymer performance, e.g. loss of mechanical strength or changes in electrical properties, should be determined by measurement of that specific property which will cause eventual failure once specified limits are exceeded. Unfortunately, mechanical and electrical properties do not change at the same rate as chemical reactions take place in a polymer. For example, a polymer which undergoes chain scission as the predominant degradation reaction will show a rapid decrease in modulus and in ultimate strength. Polymers which tend to cross-link will undergo a decrease in tensile strength and elongation. Electrical properties are dependent on reactions that form polar groups, e.g. carbonyls, in polymer molecules. The accumulation of carbonyl groups has a lesser effect on mechanical properties. It is important therefore to determine how a polymer will fail in service and to follow changes in that property which is critical to failure.

Product analysis is a laborious procedure, usually requiring elaborate apparatus and a high level of analytical skill. In oxidative degradation, the various products that could be measured analytically form as the polymer reacts with oxygen. Thus the determination of the rate at which a polymer absorbs oxygen can provide an overall measurement of reactions that lead to degradation. Oxygen uptake is often used to follow oxidative degradation. Once the relationship has been established between the rate of reaction with oxygen and a critical mechanical or electrical property in a polymer, oxygen uptake provides a simple method for continuously measuring degradation. Product analysis by nondestructive tests, e.g. carbonyl determination by infrared spectroscopy, requires interruption of the test. In destructive tests, many identical samples must be used so that one or more can be withdrawn at each time interval selected for measurement. Also many mechanical tests give scattered results due to sample variation, and multiple samples are required to obtain an average result.

In those applications where the development of color or loss in clarity are responsible for failure, comparison with color standards is often used. Very accurate results can be obtained by measurement of color development or loss in clarity using color comparison devices. Other tests for degradation are based on weight loss as measured by thermal gravimetric analysis (TGA). The exotherms that occur as a stabilizer becomes exhausted and uninhibited degradation begins form the basis for still other test methods.

III) Test for Thermal Degradation

Measurements of stability to heat are conducted either in the absence or presence of oxygen. In certain applications, e.g. heat resistant polymers, it is important to know the maximum temperature that can be sustained in the absence of oxygen or other reactants. Heat resistance as well as resistance to radiation in an inert atmosphere are important for materials that are to be used in extra-terrestrial applications. Although oxygen is present in trace amounts during conventional fabrication of polymers, nonoxidative thermal degradation (pyrolysis) can be the primary reaction responsible for failure during processing. The loss of hydrogen chloride during processing of poly(-vinyl chloride) is a typical example of thermal degradation that can occur in the absence of oxygen.

Test procedures which measure the stability of polymers against thermal oxidation are widely used to estimate the useful life of polymers and to rate antioxidants. Oxidative degradation takes place in many polymers at ambient temperatures, and at much faster rates at elevated temperatures. Because of the excellent resistance to oxidation of many stabilized polymer compositions, it is usually necessary to employ accelerated tests to determine their oxidative stability.

1) Methods for Measuring Nonoxidative Thermal Oxidation

Test designed to measure nonoxidative thermal degradation should ideally be conducted in total vacuum or under an inert gas containing no oxygen. It is difficult to achieve either of these conditions. However, oxygen can be essentially excluded by high-vacuum technology and the oxygen content of inert gases can be reduced to a negligible level. Meaningful data on nonoxidative degradation can be obtained using carefully controlled conditions. Traces of oxygen that are present will induce limited oxidation, but they are not sufficient to mask the major reactions of nonoxidative, thermal degradation.

Thermal gravimetric analysis (TGA), conducted in a high vacuum or in an inert atmosphere is widely used as a test for measuring the stability of polymers to heat. TGA measures weight loss either isothermally or at a programmed temperature increase. A highly-sensitive quartz balance is used to determine the small weight changes that occur [9]. These tests must be made under carefully-controlled conditions because very small samples of the polymer under examination are used. Products of pyrolysis can be trapped and identified as they exit from the outlet port.

Thermal volatilization analysis (TVA) is a similar test in which the pressure developed by volatile products is measured [10]. A Pirani gauge is used to give an accurate measurement of small pressure changes. Both TGA and TVA analyses are affected by the rate of heating. It is therefore important for comparative analyses to establish a standard heating rate.

Several calorimetric methods have been adapted to the measurement of nonoxidative stability. These tests detect exothermic reactions resulting from bond rupture in a polymer. Differential scanning calorimetry (DSC) [11] is frequently used to measure the resistance of polymers to heat. Differential thermal analysis (DTA) [12, 13] has also been used to study polymer pyrolysis, but since this method is more frequently used in testing for oxidative stability, it will be described in the following section.

2) Test Methods for Measuring Thermal Oxidation

Tests for measuring the rate of oxidation originated in the early studies on the deterioration of natural rubber. Many of these tests are now used to measure the stability of most of the common polymers of today. Tests have been developed to measure stability under processing conditions while others are designed to evaluate the long-term degradation that takes place under use conditions. Tests that simulate processing conditions are run at higher temperatures, and both heat and stress are applied to the polymer.

a) Tests for Measuring Stability During Processing

The milling test for determining the oxidative stability of natural rubber was perhaps the first procedure to be used for measuring the rate of degradation. This test is ideally suited to evaluating stability under processing conditions since milling is a general procedure applied in the processing of elastomers. In this test the elastomer is passed continuously between heated rolls, and failure is defined as the time when sticking to the rolls is first observed. Several modifications of the original test have evolved, and a standard test procedure has been established by ASTM [14]. Though milling tests are not very precise, reasonable correlations have been reported [15] between the time to sticking and changes in chemical composition such as the rate of carbonyl formation. Relatively high temperatures are employed in milling tests. As a result, unrealistic loss of antioxidants may occur, leading to erroneous conclusions. Loss of antioxidants would be expected to take place more rapidly than in the enclosed equipment used in the fabrication of most polymers. Milling tests are best suited for measuring processing stability although they have been used on occasion to estimate long-term aging.

The torque rheometer is an adaptation of the milling test but provides more accurate data. Several commercial rheometers are available. Most of these devices use counter-rotating blades or knives which turn at a constant speed to introduce shear as the sample is heated in the test chamber. The initial torque is recorded, and as degradation begins, an abrupt change in torque is observed. In those polymers that degrade primarily by chain scission, the torque decreases. Polymers that cross-link predominately, exhibit an increase in torque. Since this test is conducted in an enclosed chamber, antioxidant loss is restricted, and the sample can also be degraded in a controlled environment. Processing conditions are simulated by minimizing the oxygen concentration within the test chamber. In the rating of antioxidants, the test is usually run in air or in an atmosphere of high oxygen content. Because failure is measured instrumentally, the torque rheometer provides a more precise measurement of stability than does the milling test.

b) Tests for Measuring Long-term Stability

Oven-aging tests are used to measure the rate of thermal oxidation under accelerated conditions. The data obtained are then extrapolated back to actual exposure conditions. There are several versions of the oven-aging test, most of which owe their popularity to the relatively simple procedures involved. This very simplicity, however, is offset by serious shortcomings in each version of this test. Both static and air-circulating ovens are used. When an oven without air circulation is employed, additives and antioxidants in particular, vaporize into the oven cavity, and may then migrate back to contaminate other samples. For example, a control sample containing no antioxidant might become slightly stabilized by absorption of antioxidants from stabilized compositions exposed simultaneously for in previous tests. In forced-draft ovens, there is a rapid replacement of air above the test samples which could lead to an abnormal loss of antioxidants as they diffuse to the polymer surface. Even very high-molecular-weight antioxidants have vapor pressures sufficient to cause excessive loss in a flowing air stream.

In oven-aging tests, samples must be removed at selected intervals so that property changes can be measured. Any one of the property measurements discussed in Sect. G-II may be used to follow degradation in an oven-aging test. If the test which is selected gives scattered results, as for example with destructive mechanical tests, a large number of samples must be used for each time interval. Unless the approximate time to failure can be estimated, the test may have to be interrupted many times for property measurements. It is quite conceivable that all samples might be used for intermediate testing before the point of failure is reached. Samples tested by nondestructive procedures can be returned to the test chamber. There is the risk, however, that interruption of the exposure could cause a break in what should have been a continuous rate curve for degradation. Despite these problems, oven-aging tests can still provide acceptable data with reasonable accuracy if care is used to minimize or control the factors that could lead to erroneous results.

Oven-aging tests are used to measure degradation under accelerated conditions. Samples are exposed to temperatures considerably higher than would be encountered under conditions of actual use. Extrapolation of this accelerated data could lead to error in estimating the time to failure as indicated in Fig. G-1. The test should be run at several temperatures with the lowest approaching that at which the material will be used – as closely as would be practical.

Oxygen-uptake tests [16–20] are a modification of oven-aging. Many of the potential errors in oven-aging are avoided, but other errors are inherent in oxygen-uptake measurements. Ovens, baths and metal blocks have been used to heat test samples. Metal blocks provide the most uniform temperature throughout the test chamber, and this is essential when multiple samples are to be tested. A typical oxygen-uptake apparatus employing an aluminum block is shown schematically in Fig. G-2. This apparatus is readily adaptable to testing of many samples simultaneously. Each sample is placed in an enclosed system thus preventing cross-contamination. Sample tubes are filled with oxygen and are rapidly connected to measuring burettes that are also filled with oxygen. Once the test samples reach the test temperature, atmospheric pressure is established with the mercury leveling tube. Reaction with oxygen proceeds slowly when there is an effective antioxidant in the polymer. Once the antioxidant is exhausted, there is an increase in the rate of oxygen absorption as shown in Fig. G-3, eventually reaching an apparent steady rate. Since sample size is small in this test, it is usually advisable to run each sample in duplicate.

Failure in oxygen-uptake tests is usually taken as the intercept of the steady state rate with the time axis, referred to as the induction period. However, only the rate at which oxygen reacts with a polymer is measured in these tests. As previously pointed out, failure in service is determined by the deterioration of aesthetic, mechanical or electrical properties, and does not always coincide with the onset of rapid oxidation. It is quite possible that for certain applications a critical property may exceed specified limits while the polymer is still in the induction period. If the maximum amount of oxy-

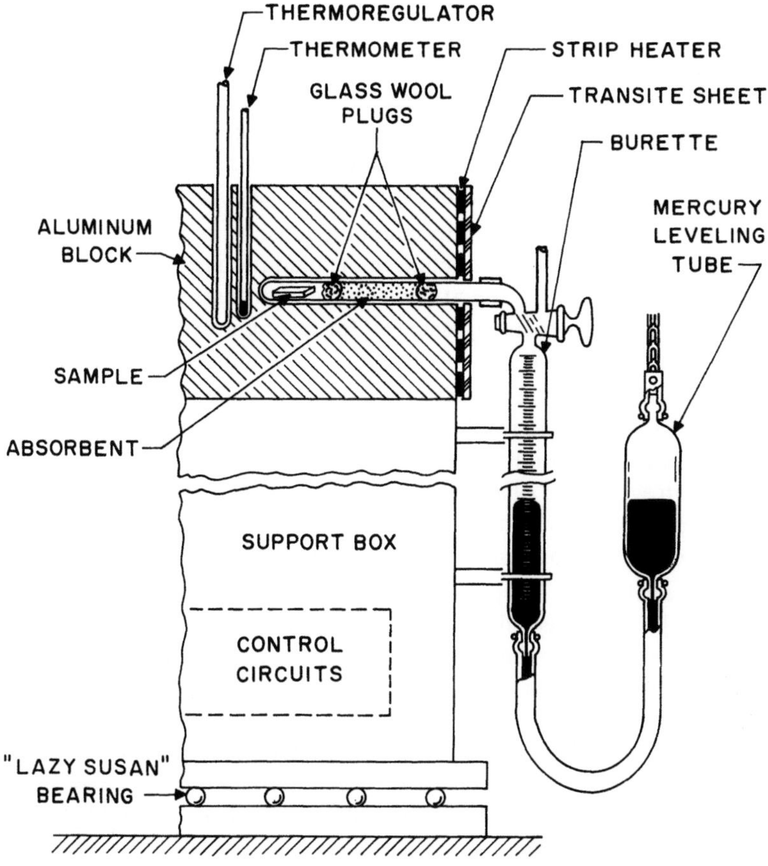

Fig. G-2. Schematic of a typical oxygen-uptake apparatus. (Reprinted with permission of Wiley-Interscience)

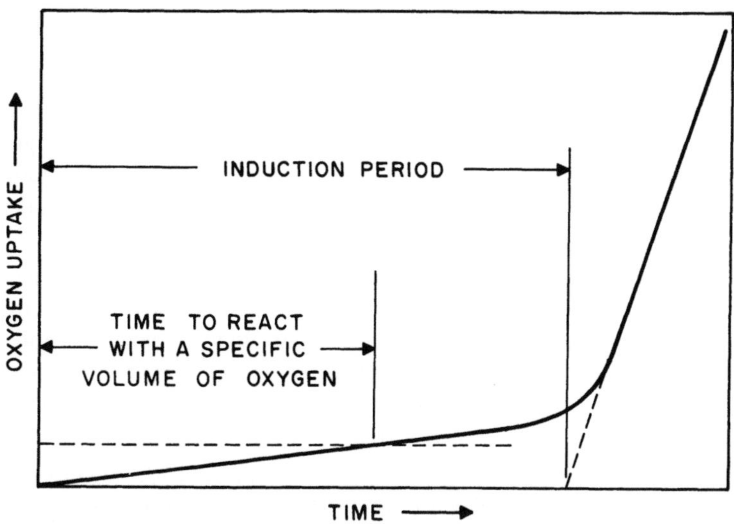

Fig. G-3. Schematic representation of the oxidation rate curve for an inhibited polymer. (Reprinted with permission of Wiley-Interscience)

gen that can react before unacceptable change in a key property takes place, then the time to failure can be taken when this level of oxidation is reached (Fig. G-3). When oxygen-uptake testing is performed in this way, true failure can be predicted as it would occur in an actual application. As with oven-aging, oxygen-uptake tests are usually run at several temperatures, and the same errors of extrapolation can occur.

Oxygen is used rather than air in order to maintain constant pressure throughout the test. Testing in air could result in a reduction in the oxidation rate as the measurements may also be controlled by the rate at which oxygen diffuses into the polymer. A sample thickness of 5 mils has been suggested [17] for testing polyolefins in order to avoid a diffusion-controlled reaction.

Loss of antioxidants is restricted in the oxygen-uptake test since samples are enclosed in small reaction tubes. This could result in a higher level of retention of antioxidant than would occur under normal conditions of use. The time to failure might then appear to be greater than would be observed under conditions that do not suppress antioxidant loss. Oxygen-uptake is primarily a materials test, but samples of laminates or insulated wire can be used thus bringing the test closer to a design test.

Differential thermal analysis (DTA) provides a rapid method for measuring stability to thermal oxidation. This test is usually run at temperatures of 200 °C or higher. Results are obtained in minutes rather than hours or days as required for most other test procedures. Typically, only a single sample is run at one time although equipment has been developed that can handle multiple samples [21]. Although the high temperature range used in DTA could result in subtle differences in reactions from those contributing a low-temperature oxidation, reasonably good correlation has been reported [22] between DTA and oxygen-uptake data. However, antioxidant loss under DTA conditions may lead to anomalous results.

A typical DTA apparatus is shown in Fig. G-4. A test sample and a reference material of the same size and geometry and that is stable under test conditions are placed in separate dishes, each with a thermocouple attached. This assembly is contained in a heated chamber with gas inlets under each dish and an outlet port. The chamber is filled with an inert gas until the test temperature is reached. Oxygen is then introduced to replace the inert gas. This change in the surrounding atmosphere causes a slight endotherm as the data is plotted, and this indicates the point at which the test begins. A strong exotherm is observed when the sample begins to oxidize rapidly. These changes are plotted as the differential temperature between sample and control. Data obtained in DTA measurements resemble those shown previously (Fig. G-3) for oxygen-uptake tests. Well-defined induction periods are observed with stabilized polymers.

DTA has the advantage of rapid measurement, but this must be balanced against the error potential in high-temperature testing. Unless a correlation has been established between DTA and lower temperature tests for the particular polymer (or polymer composition) to be tested, DTA is perhaps best suited as a screening test. It has been widely used in the screening of antioxidants [21], but care must be exercised to avoid eliminating a candidate that failed because of an unrealistic loss of the antioxidant.

Only those tests that have been used widely in practice have been included in this section on test procedures. Others have been developed for special applications but these are too numerous for inclusion in this review. Many of these are design tests in which a device is exposed to elevated temperatures and its operation monitored with time to the point of failure. Details on tests of this type can be found in literature on the particular device of interest.

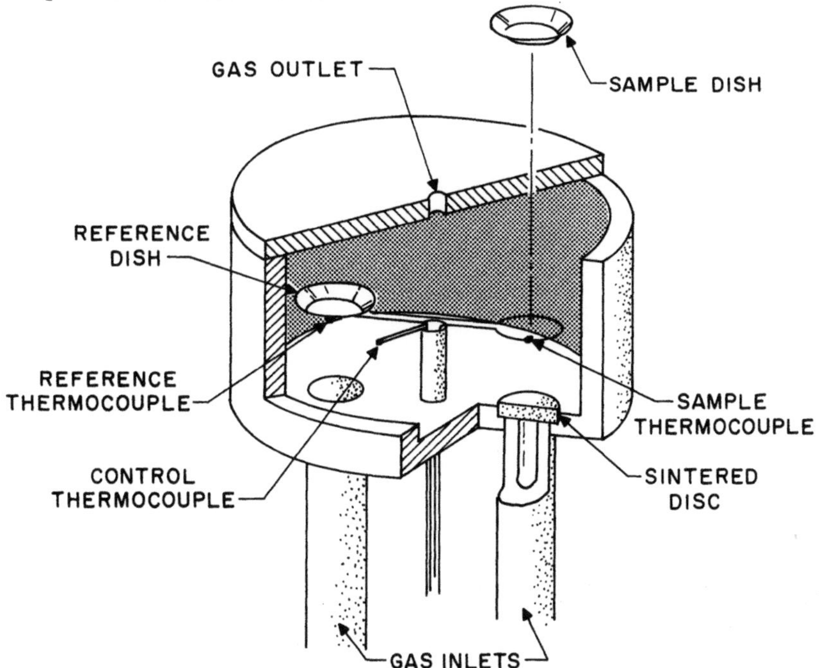

Fig. G-4. Schematic of differential thermal analysis. (From paper by E. Kokta, presented at the 24th Annual International Wire and Cable Symposium)

There has been some interest in chemiluminescence as a method for measuring thermal oxidation. This approach is based on light emission, as fluorescence or phosphorescence, that results from chemical reactions occurring in certain polymers. Chemiluminescence has not as yet found general application, but is of continuing interest as a research tool for studying polymer degradation.

There is another quite different approach to accelerated testing that although in the conceptual phase still warrants mentioning in this review. Rather than intensifying exposure conditions, this approach would seek early detection of incipient degradation as it occurs at or near the temperature at which the polymer will be used. Obviously this approach requires highly sensitive measurements. Furthermore, degradation reactions must be unterstood much more completely. It would be necessary to identify the initial reaction or product of degradation. Then the reate at which this reaction progresses must be known. Early detection methods may well be very important in the future, but much basic research and analytical technology must evolve before this approach can become a reality.

IV) Weatherability Tests

As the term implies, weatherability tests are intended to evaluate polymer performance in an outdoor environment. Degradation during outdoor exposure is influenced to varying degrees by all of the phenomena associated with natural weather conditions.

Heat, radiation (ultraviolet and infrared), rain and atmospheric contaminants all contribute to degradation of polymers when exposed out-of-doors. None of these phenomena are constant in one location, and the wide variations with location are obvious. Thus the testing of outdoor stability presents many complex problems. Any single phenomenon, e.g. uv radiation, can be maximized in the selection of an exposure site, but other contributing factors may not be intensified to the same degree. To attain the maximum accuracy in predicting the useful life of a polymer out-of-doors, all components of the real exposure environment should be considered. This can best be accomplished by conducting the test in an actual outdoor environment.

1) Outdoor Weathering Tests

Weatherability tests conducted at selected sites have provided much useful data on the weatherability of polymers. Site selection can provide environments of high humidity or hot, dry conditions. Ideally, the exposure site selected should approximate the conditions to which the polymer will be subjected in the intended application. Often this basic concept is overlooked in selecting exposure sites to provide high levels of solar radiation. If reactions with water contribute significantly to the degradation of a polymer, failure would be expected to occur sooner in areas of high humidity. When a polymer is to be used in areas having widely different weather conditions, more than one test site may be required. The major difficulty inherent in outdoor testing is the interruption in solar radiation during the night. Because of this dark interval, it is seldom practical to reach the failure point with well-stabilized formulations in outdoor tests. Nonetheless, outdoor weathering tests are used extensively, particularly in design testing since they do reflect the actual conditions of use.

Standards have been established by the American Society of Testing and Materials (ASTM) [24] for outdoor exposure sites. The angle of exposure for polymer samples has been set at 45° facing due south. Other variables including construction of the frames on which samples are mounted and the nature of the surface beneath frames [25] have also been standardized. Adoption of these standards and proper selection of the exposure site have made possible reasonably good comparative test results. However, an acceleration factor of only two or less can be realized by testing in a high uv intensity area over the time required to complete the test in a temperate region.

Several devices have been developed which rotate test samples to maximize exposure to solar radiation [26]. The acceleration factor is increased by solar tracking, but not to a great extent. In an extension of solar tracking, Caryl and Helmick [27] have developed devices that have both an equatorial mount and also a series of polished metal mirrors that accumulate and focus solar radiation onto test samples. A schematic of one of these devices is shown in Fig. G-5. Because of the intense radiation concentrated on the samples, air circulation or a water spray must be used to reduce the contribution of thermal degradation. The acceleration factor in these devices is claimed to be between 8 and 11 [28]. Although correlations with normal outdoor tests have been reported [29] for some polymer compositions, anomolies have also been observed [30, 31].

The principal objection to outdoor weathering tests is the time required to obtain results as a result of the day-night cycle. Inadequate control over the exposure environment has been criticized in these tests, but weather conditions do vary in the real en-

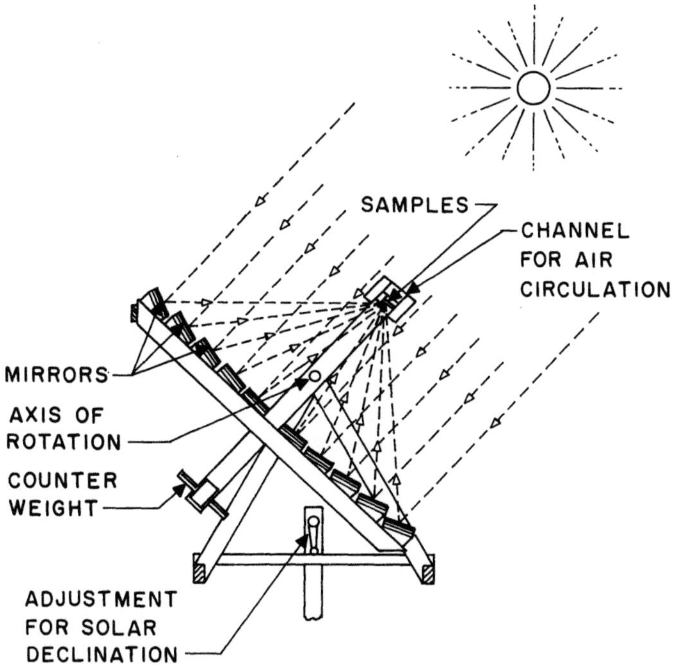

Fig. G-5. Schematic representation of an accelerated outdoor weathering device. (Reprinted with permission of Wiley-Interscience)

vironment. Indoor weatherability tests have been developed as an approach to better control over exposure conditions and to permit continuous exposure to the radiation source.

2) Indoor Weatherability Tests

Many devices or weatherometers have been designed for indoor weatherability testing. Although these devices are now quite complex, the basic principles of weatherometers can be described quite adequately using the simple apparatus shown schematically in Fig. G-6. In this device, test samples are mounted on a metal frame that rotates slowly around the source of uv radiation. With accurate frame design, the exposure of all samples is the same despite any directional variation in output from the light source. The light source used in this apparatus was an open carbon arc. Fortuitously, when applied to the testing of polyethylene formulations, this simple device gave a good correlation with outdoor testing [31] for this particular polymer. Many variables, intrinsic to this device would probably lead to a different correlation for other polymers.

The light source used in a weatherometer is very important in establishing an accurate correlation with outdoor weathering. It is most important for general application that the artificial radiation source closely resemble the sun's spectrum. Furthermore, considerable heat is generated by the open carbon arc. For this reason, the device shown in Fig. G-6 has incorporated in the design a baffle to shield the samples from heat during a portion of the cycle. Though present knowledge of the degradation process seems to rule out the need for a day-night cycle in which reverse reactions could

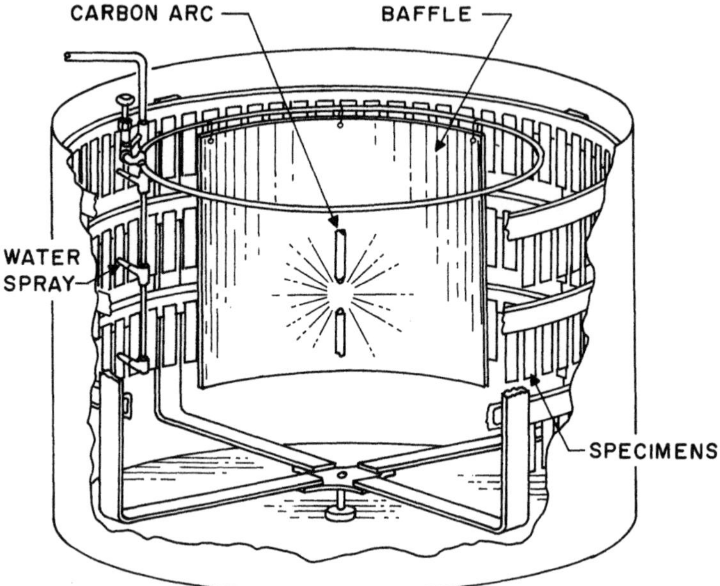

Fig. G-6. Schematic representation of an indoor weathering device. (Reprinted with permission of The American Chemical Society)

take place, this device did provide this option. Actually the baffle served only as a means of reducing sample temperature.

Cooling of the test samples by a water spray is another option incorporated into the device. Though the water spray could have been considered as a simulation of rainfall, it actually provided another means of reducing sample temperature. No attempt was made to adjust the water spray to approximate rainfall. Modern weatherometers have been designed [32)] in which moisture is condensed on the surface of test samples. This modification essentially eliminates the problem of deposits collecting on the surface as would be expected if tap water were used in a spray. Distilled or deionized water can also be used in a spray to minimize deposits on the surface of samples.

Many different light sources have been used in commercial weatherometers. In addition to the carbon arc, mercury arcs, fluorescent tubes, and xenon arcs have been used to simulate sunlight. Of these, the xenon arc emits a spectrum most closely resembling that of the sun. Devices using a xenon arc are referred to as Xenostats. Xenon arcs have a rapid decrease in emission when first put into operation. Preaging is customarily used to offset this loss.

A downward drift in output with any uv source could result in a longer, apparent time to failure than would have been observed at a constant level of radiation. Several weatherometers use an actimeter to accurately measure the output of the light source. By placing an actimeter in one of the positions normally occupied by a test sample, it is then possible to record the amount of incident radiation reaching all samples. Degradation can be based on radiation units rather than time of exposure. It would then be possible to determine from weather bureau records the time interval in which sufficient radiation to cause degradation would reach the earth's surface at any selected location. The progress of polymer degradation during indoor testing could be monitored by measurement of any aesthetic, mechanical or electrical property described in Sect. G-II.

Most indoor weathering tests do not need the level of sophistication described above in order to provide meaningful data. A less precise measurement would be acceptable in many cases if a factor of safety is used. In other words, a reasonable over design in polymer stability might be more practical than application of an elaborate testing procedure.

V) Miscellaneous Test Procedures

The accelerated tests for measuring the degradation of polymers that have been discussed up to this point have used intensified exposure conditions to reduce the time required for testing. Acceleration can also be accomplished by exposing a polymer to an environment in which there is an abnormally high concentration of the chemical reactant responsible for degradation. Measurement of the resistance of stressed elastomers to ozone-induced degradation is an example of this approach to accelerated testing.

Crabtree and Kemp[33] developed a method for measuring the stability of stressed elastomers to ozone attack. In this test, samples are exposed to 50 pphm of ozone in a stream of oxygen and at temperatures between 40 and 50 °C. This concentration of ozone is considerably higher than the normal range of 0 to 10 pphm found at most locations on the earth's surface. It is, however, comparable to that recorded on occasion at highly poluted locations. The temperatures used in this test are also slightly higher than those encountered in most applications. The higher temperature contributes to acceleration, but not to the extent resulting from the increased ozone concentration.

Standardized conditions have been established by the American Society for Testing and Materials (ASTM) for measuring ozone resistance[34]. Failure of an elastomer in this test is observed visually as the point at which cracks first appear. Test samples are clamped in a mandrel of fixed dimensions to provide a uniform stress[35]. Very careful control of the ozone concentration is essential in this test. Both moisture and uv radiation can influence the rate of ozone-induced degradation occurring under normal conditions of use, and neither of these factors is taken into account in the more commonly used procedures.

Although tests based on high ozone concentration are used extensively to evaluate antiozonants, results from these tests can be misleading for reasons discussed in Sect. F-II. According to the most widely accepted mechanisms for antiozonant protection, these additives must migrate to the elastomer surface in sufficient amount to be involved in reactions that modify the stressed surface. Initial exposure to high ozone concentrations may not allow sufficient time for migration and the subsequent reactions to take place. It is quite possible for an antiozonant to perform poorly in an accelerated test and yet this same additive could be quite effective under normal exposure conditions. When an elastomer is to be used under dynamic conditions, as in automobile tires, the test procedure is much more complicated. Standards have not been established for dynamic testing except as they apply to specific products.

Polymers that degrade as a result of hydrolytic reactions can be tested under high humidity conditions in order to accelerate the time to failure. Weight gain under controlled humidity conditions is widely used. The permeability to moisture has also been employed [36]. The rate at which molecules are cleaved by hydrolysis can be followed by changes in viscosity [37]. Acid or base catalysis will accelerate hydrolysis reactions, but the extent of catalysis should at least approximate that to be encountered under normal use conditions.

Although mechanisms for the burning of polymers has not been discussed in this review, a word about testing for flammability will serve to emphasize many of the potential errors inherent in accelerated or laboratory test procedures. Many attempts have been made to develop tests that correlate with real fires [38]. Lack of an acceptable correlation invariably occurs when anyone of several important factors that contribute to an actual fire are not incorporated into the laboratory test. For example, the effect of radiant heat in a real fire has often been overlooked.

Whether it be in burning, thermal oxidation, weathering or hydrolysis, an accelerated test should take into account every factor that contributes to degradation under the conditions to which the polymer will be exposed in actual use. Furthermore, the contribution of each of these variable factors must be proportional to that existing in the real environment. For example, accelerated tests for outdoor weathering that cause abnormal heating of test samples could actually be measuring thermal degradation as a major factor contributing to failure when this mechanism might have made a negligible contribution under normal exposure conditions.

An unequivical prediction of a polymer's useful life can only be obtained by incorporating the polymer into the device in which it will be a component part, and then exposing the finished product to the actual environment in which it is to be used. The best accelerated tests available only approximate the result obtained in this way. Nonetheless, accelerated tests can provide meaningful data when properly designed. Testing under use conditions simply requires too much time to be practical in most instances.

References

1. Kelen, T.: Polymer Degradation, p. 40, New York: Van Nostrand Reinhold 1983
2. Tamblyn, J.W., Newland, G.C., Watson, M.T.: Plastics Technol., *4(5)*, 427 (1958)
3. Gedemer, T.J.: Appl. Spectroscopy, *19(5)*, 141 (1964)
4. Hearst, P.J.: Amer. Chem. Soc. Polym. Preprints, *28(1)*, 672 (1968)
5. Chan, M.G., Hawkins, W.L.: Amer. Chem. Soc. Polym. Preprints, *9(2)*, 1638 (1968)
6. Kamal, M.R., Saxton, R.: Recent Developments in the Analysis and Prediction of the Weatherability of Plastics, in: Weatherability of Plastic Materials (ed) Kamal, M.R., p. 9, Appl. Polym. Symposia No. 4, New York: Wiley 1967
7. Bruck, S.D.: Polym., *7*, 321 (1966)
8. Schard, M.P., Russel, C.A.: J. Appl. Polym. Sci., *8*, 985 (1964)
9. Madorsky, S.L.: Thermal Degradation of Organic Polymers, p. 10, New York: Wiley 1964
10. McNeill, I.C.: J. Polym. Sci., Part A-1, *4*, 2478 (1966)
11. Ellerstein, S.M.: The Use of Dynamic Differential Calorimetry for Ascertaining the Thermal Stability of Polymers, in: Analytical Calorimetry (eds) Porter, R.S., Johnson, J.F., p. 279, New York: Plenum 1968
12. Rodin, A., Schreiber, H.P., Waldman, M.H.: Ind. Eng. Chem., *53(2)*, 137 (1961)
13. Stafford, B.B.: J. Appl. Polym. Sci., *9*, 729 (1965)
14. ASTM Book of Standards, D 1248, *26*, 79 (1969)
15. Baum, B., Perun, A.L.: Plastics Technol., *7(4)*, 29 (1961)
16. ASTM Book of Standards, D 525, *17*, 198 (1969)
17. Biggs, B.S., Hawkins, W.L.: Mod. Plastics, *31*, 121 (1953)
18. Shelton, J.R., Winn, H.: Ind. Eng. Chem., *38*, 71 (1946)

19. Shelton, J.R.: Amer. Soc. Testing Mater., Spec. Tech. Publ., *89*, 12 (1949)
20. Wilson, J.E.: Ind. Eng. Chem., *47*, 2201 (1955)
21. Howard, J.B.: Polym. Eng. Sci., *13*, 429 (1973)
22. Howard, J.B.: unpublished results
23. Schard, M.P., Russell, C.A.: J. Appl. Polym. Sci., *8*, 985 (1964)
24. ASTM Book of Standards, D 1435, *27*, 509 (1969)
25. Rugger, G.R.: Weathering, in: Environmental Effects on Polymeric Materials Vol. I (eds) Rosato, D.V., Schwartz, R.T., p. 350, New York: Interscience 1968
26. Gardner, H.A., Sward, G.G.: Paint Testing Manual, 12th ed., p. 262, 1962
27. Caryl, C.R., Helmick, W.E.: US Patent No. 2,954,417 July 19, 1960
28. Papillo, P.J.: J. Paint Technol., *40(524)*, 359 (1968)
29. Kennedy, R.J, Beardsley, H.P.: "Exterior Paints for Wood Based on Vinylethylene Emulsions", presented at the Annual Symposium of the Northwest Society of Paint Technology, May 6, 1967
30. Martinovich, R.J., Hill, G.R.: Practical Approach to the Study of Polyolefin Weatherability, in: Weatherability of Plastic Materials (ed) Kamal, M.R., p. 144, New York: Interscience 1967
31. Howard, J.B., Gilroy, H.M.: Polym. Eng. Sci., *9*, 286 (1969)
32. Grossman, P.R.: ASTM STP 646, p. 5, 1978
33. Crabtree, J., Kemp, A.R.: Ind. Eng. Chem., Anal. Ed., *18*, 769 (1946)
34. ASTM Book of Standards, D 1149, *28*, 563 (1969)
35. ASTM Book of Standards, D 518, *28*, 300 (1969)
36. Stannett, V., Yasuda, H.: Permeability, in: Crystalline Olefin Polymers (eds) Raff, R.A.V., Doak, K.W., p. 131, New York: Interscience 1964
37. McMahon, W. et al.: J. Chem. Eng. Data, *4*, 57 (1959)
38. Hilado, C.J.: Flammability Handbook for Plastics, Technomic, Westport, Connecticut 1969
39. Rowen, J.V., Lyons, J.W.: J. Cellular Plastics, 25 (1978)

H) Future Trends

A prediction of future developments in polymer degradation and stabilization can be approached by consideration of those areas wherein a deficiency in basic knowledge and/or stabilizer development is perceived. There are some areas in which only a steady development in new products should be expected. In other areas where application of polymers has and continues to be restricted by the inability to assure an adequate useful life, major breakthroughs should be expected.

Protection against thermal oxidation has been the object of concerted research for several decades. As a result, antioxidants have been developed which provide adequate protection for most current applications. Design requirements, however, continue to be more demanding as new applications for polymers are found. Recent developments in antioxidants have been oriented toward additives having greater retention in polymers. Still higher molecular weight antioxidants may yet be required for greater permanence, particularly at high temperatures. The extended use of polymers in automobile manufacture will probably provide some of the impetus to development of a new generation of antioxidants. One area of interest will be in antioxidants that can be used in the reaction-injection-molding (RIM) process. This process is of considerable interest for the fabrication of large parts. New polymers for the RIM process are being developed and these may well require stabilizer systems designed specifically for them.

An unacceptable rate of photodegradation continues to limit the use of polymers in many applications. The introduction of hindered amine light stabilizers has been a major advance in the protection of polymers against photodegradation, but much remains to be learned about how these additives function. Contributions of photochemists will continue to supplement the research of polymer chemists in elucidating the mechanism responsible for photodegradation. These collaborative efforts should provide the knowledge required to introduce a new generation of light stabilizers capable of protecting polymers in some of the future applications. Among these applications is the growing need for more light-stable polymers or polymer compositions to be used as encapsulants for solar cells.

Accelerated test procedures in general can be expected to improve as the need for more accurate predictions of the useful life of polymers is demanded. This is particularly likely to occur in weatherability tests. It is conceivable that artificial weathering devices will be developed which are capable of simulating any desired outdoor environment. Every variable contributing to outdoor failure might be incorporated in such devices, and computer programmed to correspond to selected weathering conditions.

It is generally agreed that the synthetic polymer industry will continue its rapid growth for quite some time. This growth will be sustained by the entrance of polymers into many new fields of application. Polymers will compete with more conventional materials in many new applications and replace many of them. For this to take place, it must be demonstrated that polymers can provide the service life demanded by design engineers. New generations of stabilizers of all types will certainly emerge to meet this challenge.

Subject Index

Polymers Properties and Applications

Editorial Board:
H.-J. Cantow, H.J. Harwood,
J.P. Kennedy, A. Ledwith,
J. Meißner, S. Okamura,
S. Henrici-Olivé, S. Olivé

Springer-Verlag
Berlin
Heidelberg
New York
Tokyo

Crazing in Polymers

Editor: **H. H. Kausch**
With contributions by numerous experts

1983. 234 figures, 12 tables. Approx. 390 pages
(Advances in Polymer Science, Fortschritte der Hoch-
polymeren-Forschung, Volume 52/53)
ISBN 3-540-12571-X

Contents/Information:

E. J. Kramer: **Microscopic and Molecular Fundamentals of Crazing.** The first chapter considers crazes produced in air. The electron beam imaging technique developed in Ithaca is shown to be a powerful tool to determine the micromechanics and microstructure of crazes. (133 ref.)

M. Dettenmaier: **Intrinsic Crazes in Polycarbonate: Phenomenology and Molecular Interpretation of a New Phenomenon.** The second chapter demonstrates that the intrinsic crazing of polycarbonate is related to the existence of a general mode of cavitational plasticity in glassy polymers. (193 ref.)

W. Döll: **Optical Interference Measurements and Fracture Mechanics Analysis of Crazes.** The third chapter describes the optical interference measurement providing considerable insight into the role of crazes in deformation and fracture of amorphous polymers. (142 ref.)

J. A. Sauer, C. C. Chen: **Crazing and Fatigue Behavior in One- and Two-Phase Glassy Polymers.** The fourth chapter investigates deformation and fracture mechanisms in glassy polymer systems subject to monotonic loading. (74 ref.)

K. Friedrich: **Crazes and Shear Bands in Semi-Crystalline Thermoplastics.** The fifth chapter reviews the interaction between the microstructure of crystalline polymers and the craze or shear band formation at various temperature and loading conditions. (151 ref.)

A. S. Argon, R. E. Cohen, O. S. Gebizlioglu, C. Schwier:

Crazing in Block Copolymers and Blends. The last chapter reports on the principles of polymer toughening by crazing in block copolymers and blends. (68 ref.)

Springer-Verlag
Berlin
Heidelberg
New York
Tokyo